T0184555

What is Global Engineering Education For?

The Making of International Educators

Part III

Synthesis Lectures on Global Engineering

Editor
Gary Downey, *Virginia Tech*
Assistant Editor
Kacey Beddoes, *Virginia Tech*

The Global Engineering Series challenges engineering students and working engineers to cross the borders of countries, and it follows those who do. Engineers and engineering have grown up within countries. The visions engineers have had of themselves, their knowledge, and their service have varied dramatically over time and across territorial spaces. Engineers now follow diasporas of industrial corporations, NGOs, and other transnational employers across the planet. To what extent do engineers carry their countries with them? What are key sites of encounters among engineers and non-engineers across the borders of countries? What is at stake when engineers encounter others who understand their knowledge, objectives, work, and identities differently? What is engineering now for? What are engineers now for? The Series invites short manuscripts making visible the experiences of engineers and engineering students across the borders of countries. Possible topics include engineers in and out of countries, physical mobility and travel, virtual mobility and travel, geo-spatial distributions of work, international education, international work environments, transnational identities and identity issues, transnational organizations, research collaborations, global normativities, and encounters among engineers and non-engineers across country borders. The Series juxtaposes contributions from distinct disciplinary, analytical, and geographical perspectives to encourage readers to look beyond familiar intellectual and geographical boundaries for insight and guidance. Holding paramount the goal of high-quality scholarship, it offers learning resources to engineering students and working engineers crossing the borders of countries. Its commitment is to help them improve engineering work through critical self-analysis and listening.

What is Global Engineering Education For? The Making of International Educators - Part III
Gary Lee Downey and Kacey Beddoes
2011

What is Global Engineering Education For? The Making of International Educators - Part III

Gary Lee Downey and Kacey Beddoes

ISBN: 978-3-031-00997-6   paperback
ISBN: 978-3-031-02125-1   ebook

DOI: 10.1007/978-3-031-02125-1

A Publication in the Springer series
SYNTHESIS LECTURES ON GLOBAL ENGINEERING

Lecture #2
Series Editor: Gary Downey, *Virginia Tech*
Assistant Editor: Kacey Beddoes, *Virginia Tech*
Series ISSN
Synthesis Lectures on Global Engineering
ISSN pending.

What is Global Engineering Education For?

The Making of International Educators

Part III

Gary Lee Downey and Kacey Beddoes

Virginia Tech

SYNTHESIS LECTURES ON GLOBAL ENGINEERING #2

ABSTRACT

Global engineering offers the seductive image of engineers figuring out how to optimize work through collaboration and mobility. Its biggest challenge to engineers, however, is more fundamental and difficult: to better understand what they know and value *qua* engineers and why. This volume reports an experimental effort to help sixteen engineering educators produce "personal geographies" describing what led them to make risky career commitments to international and global engineering education. The contents of their diverse trajectories stand out in extending far beyond the narrower image of producing globally-competent engineers. Their personal geographies repeatedly highlight experiences of incongruence beyond home countries that provoked them to see themselves and understand their knowledge differently. The experiences were sufficiently profound to motivate them to design educational experiences that could provoke engineering students in similar ways.

For nine engineers, gaining new international knowledge challenged assumptions that engineering work and life are limited to purely technical practices, compelling explicit attention to broader value commitments. For five non-engineers and two hybrids, gaining new international knowledge fueled ambitions to help engineering students better recognize and critically examine the broader value commitments in their work.

A background chapter examines the historical emergence of international engineering education in the United States, and an epilogue explores what it might take to integrate practices of critical self-analysis more systematically in the education and training of engineers. Two appendices and two online supplements describe the unique research process that generated these personal geographies, especially the workshop at the U.S. National Academy of Engineering in which authors were prohibited from participating in discussions of their manuscripts.

KEYWORDS

engineering education, global engineering (education), international engineering (education), study abroad, international service learning, international co-op, international internship, work abroad, Peace Corps, ABET, EC 2000, globalization

For my students, who give my work meaning

Gary Downey

For my mentors, who have given me more than they know

Kacey Beddoes

Contents

PART III

Redefining Engineering:
Five Non-Engineers and Two Hybrids

CHAPTER 12

Communicating Across Cultures: Humanities in the International Education of Engineers

Bernd Widdig

MOVING BETWEEN CULTURES

Culture, I tell my students, is essentially "the way we do things around here." I like this admittedly somewhat simple definition because it encapsulates the hidden, taken-for-granted nature of what we call "culture" – a highly complex system of a certain set of norms, values, codes of conduct and communication, narratives, and symbols. To drive this point home to my MIT students, I sometimes compared culture with a well-functioning operating system that runs in the background of our emotions, thoughts, and actions assuring compatibility with our surrounding environment. We simply live our daily life in labs and classrooms, in universities and companies, in dorms and family homes without constantly reflecting on the basic framework of our own cultural norms. It is when we step outside a certain cultural system and enter into a "foreign" system of culture that we experience those slight or at times severe compatibility problems: Why don't they function the way I do? Why don't they see things the way we do? Why is their behavior so strange…?

I begin the account of my personal and professional journey with this very brief reflection on culture and communication because the movement into another culture and the mediation between cultures has been a guiding "Leitmotiv" of my own biography and professional career. It entails the never-ending challenge to overcome those sometimes trivial, occasionally momentous compatibility problems that besiege communication between cultures, be it between Germans and Americans or between humanists and engineers. My first move into another culture came with my decision to leave Germany to pursue further graduate work at Stanford University in Palo Alto, California. During my years as a professor in German Studies at MIT, I crossed the border of the humanities and cooperated with engineers and scientists on projects in international engineering education. As Associate Director of the MIT International Science and Technology Initiative (MISTI), I learned to communicate with partners in globally operating companies. And finally, in 2007, I moved from MIT, which has an engineering and science oriented culture, to Boston College, a university with a strong tradition in the liberal arts, business, law, and education.

FROM ADENAUER ALLEE TO PALM DRIVE

I cannot claim it was my own clever planning that brought me to enroll into a Ph.D. program in German Studies in the United States in 1984. Crossroads in life often emerge through fortuitous encounters. In my case, it was my decision as a young student at the University of Bonn who minored in German literature to take a seminar with a visiting professor from Stanford University. The way he taught his class, his interaction with us as students brought a sudden ray of sun into our otherwise mostly grey and uninspiring German academic landscape. He must have sensed my light deprivation because he encouraged me to come as an exchange student to Stanford, and after I finished my German *Staatsexamen* in Economics, Political Science, and German Literature, I followed his call. There was more sun and energy on Stanford's gorgeous campus than I could ever have imagined – both inside and outside the classroom.

The considerable cultural differences I encountered between German *Germanistik* and American *German Studies* might not be a thrilling topic for professors of engineering, but the work with the faculty in the German Studies Department at Stanford gave me a profound sense of how new methodological perspectives can generate a different set of questions and, consequently, different research projects. To study my own culture from the outside, from the vantage point of the United States, has given me an ongoing intellectual pleasure over the years.

My time in California was a momentous encounter that changed almost everything in my life. For sure, Palm Drive at Stanford was a long way from the Adenauer Allee in Bonn. Being completely fluent in English was not a common characteristic of a student in the humanities in Germany in the early 1980s, so after my arrival at Stanford, I shared a house with American students to improve my English as quickly as possible. I contributed greatly to the overall entertainment of our house community when I talked about "hand shoes" (*Handschuhe* = gloves) and displayed a seeming inability to distinguish between a desert and a dessert in my Teutonic pronunciation. "Who wants desert?" became a running joke at the dinner table. The German Department at Stanford where I studied for my Ph.D. quickly became an intellectual home for me, not least because in the beginning, it provided me with a sort of enclave.

Four observations I first made as a graduate student strike me as relevant for colleagues in engineering who sometimes ask me about the culture of foreign language departments. First, foreign language departments tend to be culturally hybrid places. I am not sure about all the various reasons that attract people to study and later teach engineering. In the case of professors and teachers of foreign languages, professional and personal identities are often deeply intertwined. Many foreign language and literature professors were born abroad and immigrated to the United States; some were pressed to leave, others had sought better employment opportunities in this country. Since it is their professional task to convey the language and culture of their country of origin to young students, they personify their own educational project in their teaching. Not many outside of interdisciplinary engineering studies would associate the persona of a mechanical engineering colleague with a certain way to construct a turbine or an engine, yet when a French professor who was born and raised in France interacts with American students and faculty, he or she represents to a considerable degree "France" as a cultural and identity-defining concept. For some teachers of foreign languages, this position creates great enjoyment, for others a continuing split since their own relationship to their culture of origin may be quite complex and sometimes conflicted.

Secondly, one often observes a peculiar *Ungleichzeitigkeit* in foreign language departments, an asynchronicity of times that fosters nostalgia, that allows people to live out facets that have long vanished in the cultures they teach. It is surprising to me that even today I find German departments whose bulletin boards are still covered with flyers and welcome signs printed in *Fraktur*, an old German typescript that has long disappeared in Germany.

Third, I observed that some faculty members born in the United States connect a certain cultural or political utopia with the country of their chosen field. Highly critical and often disappointed in their own American culture or its political developments, they project onto the foreign language and culture they teach what they miss at home.

My last observation concerns graduate education in the humanities, especially in more esoteric fields such as German Studies. I became aware that by joining a Ph.D. program I had entered into something similar to a religious order. Most of us enrolled in those programs to become professors of German. There was a powerful hermetic circle that is also characteristic of a religious order. A group of older scholars teaches a younger cohort of students who will eventually become scholars in the same field and continue a venerable tradition. This creates a wonderfully intense and enriching intellectual atmosphere – one of the highlights of my graduate years. As students, we were taken seriously, we were junior members in a great intellectual order that had branches all over the country. This order gave us a home and a direction – not just for our studies, but also for the rest of our lives. There was supposed to be no doubt of our lifelong scholarly interest in German literature and culture. As a religious order does not offer alternative career paths for their members and expects a lifelong commitment, nobody in our department talked about or even conceptualized alternative pathways to a career as a professor. And for many of us, this path turned out to be our lives' direction.

Yet for those of us whose lives turned out to be different, leaving the order was not as easy as in other professions. During my Stanford days, several of us lost faith and left the order. Officially, it was seen as a sad indication of failure, even for those who took the amazing opportunities that Silicon Valley offered and put their talents into developing software or started translation companies. I suspect that this identity with a field creates a strong sense of a discipline – and, in turn, creates a desire for strong disciplinary academic borders, for clear demarcation lines between inside and outside. This might be a reason why colleagues in the humanities, and especially foreign languages, often have a hard time crossing borders despite their well-intentioned call for "interdisciplinarity." But more on this later.

Needless to say, I did not know a single engineering student at Stanford, even though many of them passed my way every day. Yet there were other encounters that shaped the way I looked at things. I got to know entrepreneurs. I lived in the heart of Silicon Valley, and the computer and software industry was experiencing its first boom in the mid 1980s. One day, I saw an advertisement in the German Department placed by a couple of software engineers who needed help with German grammar. I responded and was hired to work with them in their early attempt to develop a spell checker for word processing that included a German version. I remember their tenacity, their raw curiosity, and their breathtaking optimism. I was keenly aware that I had entered into a completely new world. The spell checker never made it to market and my two software engineering friends soon turned their attention to other, more lucrative ideas. But the more I looked outside the confines of my scholarly enclave, the more fascinated I became by the boundless energy, the "let's try" attitude of the entrepreneurial culture of Silicon Valley. And I learned something that was radically different from my German cultural codes: those who try may fail, but they are not failures.

COMMUNICATING ACROSS THE *INFINITE CORRIDOR*

In 1989, I joined MIT as an Assistant Professor in German Studies, first in a visiting position, then on the tenure track. MIT would become my professional home for the next eighteen years before I crossed the Charles River in August 2007 to become Director of the Office of International Programs at Boston College. MIT profoundly influenced my professional identity and, in ways I did not foresee, within the career of a scholar of German literature and culture. When I was going to teach a class and walked through one of the long hallways – the longest one aptly called the "Infinite Corridor" – passing along the different labs and making my way through the crowds of equally exhausted and exhilarated students, I could feel the intellectual energy and curiosity that carries this place forward. Nobody can "sail along" at MIT. The institutional culture constantly challenges its members to "stake their claim," to communicate their particular intellectual contribution to the rest of the Institute. At the core of MIT is a tremendous entrepreneurial spirit. Consequently, one's own academic identity remains more fluid and open to change than in other academic cultures.

The vast cultural difference between MIT and my academic upbringing in Germany became clear to me one day when the rector of a prominent German university visited MIT. At lunch, our German guest was asked to briefly introduce his institution. My colleagues and I then received a twenty-minute-long tour through the sometimes illustrious, sometimes dark history of his university that was deeply intertwined with the ups and downs of German history. Our visitor only elaborated the last three minutes on the current situation and the future plans of his institution. Then it was our Provost's turn. In less than two minutes, he covered the first 140 years of MIT, and for the next ten minutes, he spent on MIT's ambitious plans for the future to ensure the Institute's continued preeminence among the world's research universities.

While Stanford boasted a large and prominent German Studies Department, MIT had a Foreign Language and Literature Section in which I was the only tenure track professor in German. It is safe to say that Foreign Languages was operating at the periphery of MIT's academic culture. My senior colleagues, mostly in Spanish and French, bemoaned this fact and expressed their discontent in differing shades of anger, cynicism, and a peculiar sort of cultural elitism to which some colleagues in the humanities resort in the presence of engineering and science colleagues with large research budgets and hordes of graduate students. I brought my sunny Californian disposition into this group, despite dire warnings by my senior colleagues that nobody at MIT "respects the work we are doing here in the humanities." I soon asked myself how I could connect my work and interests with the large engineering and science community. Contrary to my colleagues, I saw our peripheral status as an advantage. We did not have to constantly raise large resources to finance graduate students and expensive research projects. We had the privilege of working with some of the most curious and knowledge-hungry young undergraduates in the country. True, they often had a dismal background in history and literature, but they very quickly asked the right questions.

During the mid 1990s, I continued research on my book *Culture and Inflation in Weimar Germany*,[1] but I also started asking myself: what is my role as a professor in the humanities in an academic environment that is primarily focused on science and engineering? How do I link my teaching to my students' aspirations as future leaders in those fields? Did it make sense to teach them about Germany when they didn't have the chance to interact with present-day Germans and contemporary German culture? As with other engineering and science-oriented universities and many Ivy League schools at the time, almost no students at MIT went abroad to study. In a larger context, the term of "applied humanities" began to intrigue me, a concept that later found entry in a different arena through MIT's successful Comparative Media Studies Program. How could one bring one's own expertise as a humanist into a creative dialogue within an engineering and science oriented community? I decided that the best way to create opportunities for our students to experience Germany's multi-layered culture in real time was to arrange internship opportunities in German companies and research labs. The idea was to establish a MIT-Germany Program so that students could combine their learning of German language and culture with a practical experience in their major field.

I remember my first contacts with German MIT alumni who were very supportive of my idea and introduced me to German companies. A key moment in this narrative is my first travel to Germany in 1996 where I had meetings with several senior officers of large German companies. Elevators brought me to the upper floors, and I now realize that it was only the MIT name that allowed me to get a time slot in the calendar of those senior managers. They were at times amused that I had come to ask them personally for the arrangement of some paid student internships. But my innocence and my entrepreneurial enthusiasm must have charmed them. In short, I was surprised and delighted by how open, how friendly, how reciprocal my "let's do this" attitude was.

I also remember one revelatory moment in those early days. When I had my appointment with Siemens, I was waiting outside a boardroom in eager anticipation of my host. The door opened and a group of about thirty people flowed into the lobby still discussing the decisions made in the meeting. Nobody among them spoke German. In fact, the language of the entire floor was English.

---

[1]Widdig, Culture and Inflation in Weimar Germany, 2001.

My host was enthusiastic about the possibility of bringing our MIT students into labs because, as he said: "Our personnel should use any opportunity to improve their English." I had not anticipated such a response and insisted that, nevertheless, all our students had to learn German before they started an internship. "Sure," he grumbled, "I see your point, they will not only work in Germany. But you have to understand, Siemens is not a German company anymore, it's a global player, and its official language is English." In a short time, I had secured twenty-five internships in several large companies. Additional funding from the German government allowed me to hire a coordinator and to start the MIT-Germany Program in 1997.

COMPATIBILITY PROBLEMS

What a different scene in my home department[2]. My "entrepreneurship" had not gone unnoticed by my senior colleagues, and only later I realized that for them this was a code word with two meanings: shoulder clapping and congratulations for the funds I had secured, but a deep misgiving that it was inappropriate for a young humanist to venture out into the business world and arrange internships in companies. Entrepreneurship was for business people and I was hired to pursue the humanistic study of German literature – not to drag myself down into the lowlands of utilitarian ends. Only later, I realized that what I naively felt was to the advantage of the whole department my senior colleagues perceived very differently. For some of them, I was "sleeping with the enemy," and their fear might have been that I could focus my attention on Business German instead of on the publication of scholarly articles.

The small group that was teaching German consisted of two lecturers, a part-time lecturer, and myself as a tenure track faculty. Enrollment figures in German had declined precipitously during the 1990s. I saw the MIT-Germany Program with its internship opportunities as a way to draw new students into our German courses, since three semesters of German were the prerequisite for applying to an internship. While enrollment in German continued to decline in many other departments around the country, we saw the reverse trend. By the end of the 1990s, more than half of those students who were taking German at MIT were in some way connected to the MIT-Germany Program. Many were taking language courses to prepare for the internships, others had come back from Germany with the desire to deepen their language competence and learn more about its culture and literature. I had the good fortune that all my colleagues in German enthusiastically supported my efforts. Within Foreign Languages and Literatures, even though we were the smallest of the language groups, the German group was for sure the most active and entrepreneurial. For example, with corporate support we instituted a yearly Lufthansa Prize for Excellence in German Studies that honored two students with round-trip tickets to Germany. With support from the Max Kade Foundation, we established the Max Kade Writer-in-Residence Program that allowed young German writers to stay at MIT for a semester and teach a course on contemporary German literature to our students.

---

[2]I use the common term "department" here. Since Foreign Languages and Literatures at MIT does not have a graduate program, it is called in MIT terminology a "section."

The resistance of my senior colleagues in the department might be confusing to an engineer who is likely to ask: what's wrong with increasing the enrollment and having students practice their foreign language skills in the real-life working environment of a host country? Since my senior colleagues' position continues to be more the norm than the exception in current foreign language teaching, let me briefly summarize their view.

For them, the learning of a foreign language is supposed to lead students to a level of mastery that allows them to read increasingly complex literary texts in the target language. At that point, those students enter into an intellectual space many professors in foreign languages identify with and for which they feel they are appropriately trained. At least for a short time, students participate in the target language in the analysis of a text by Brecht or Borges, the interpretation of a French film through psychoanalysis, or the reading of an African novel through the prism of postcolonial theory. In other words, the road of language learning does not lead into the complexities of a political, anthropological, and economic everyday culture, but rather to the concept of culture with a capital "C," the established literary and cultural canon of a foreign country.

There are several reasons for the enduring power of this pathway. The first one is the significance of a literary canon. Let me stress that I wholeheartedly support the idea of such a canon. While in a continual process of change and adaptation, cultural canons consist of works that expose students to the unique power and importance of aesthetics in a given historical and cultural context. In this sense, Goethe's works should be taught and read for their enduring literary power and beauty, and they should serve as important milestones that illuminate a specific part of German cultural history. But I think it is fundamentally a misconception to believe that reading Goethe helps students to understand the present cultural reality of Germany. To find intellectually stimulating and convincing ways to help students access and analyze a country's contemporary culture remains a challenge.

Another reason is that professors of foreign languages and literatures are usually not trained to combine language learning with a sophisticated anthropological, sociological, political, and economic approach. And finally, there are strong and complex issues of professional identity I mentioned earlier that make humanists hesitant to leave their established paths.

Throughout my years in Foreign Languages and Literatures at MIT, it was clear to me that my scholarship would be the only achievement that would count for my tenure process regardless of how many interns I placed in Germany. After publishing my first book in Germany[3], I worked hard to finish my second book project and placed it with a publisher in time for the tenure decision.[4] In the meantime, the MIT-Germany Program became a great success, and soon I found like-minded colleagues in other departments at MIT who were concerned about how we were preparing students for an increasingly international work place. One senior colleague was Richard Samuels, an accomplished political scientist whose work on Japan had led him to create a similar internship program in Japan already during the 1980s. The MIT-Japan Program was at an advanced state of development and showed me what all could be done. Another leading senior political scientist, Suzanne

---

[3]Widdig, Männerbünde und Massen, 1992.
[4]Widdig, *Culture and Inflation in Weimar Germany,* 2001.

Berger, had found her way into international education through her research on globalization. In the mid 1990s, she became the founding Director of the International Science and Technology Initiative (MISTI) and continues to be the driving force behind MISTI's continuing success. My program became part of this initiative together with the MIT-Japan Program and a newly established MIT-China Program.

In hindsight, I understand why my colleagues in Foreign Languages and Literatures decided not to make me a permanent member of their group. Though I fulfilled all scholarly expectations, I had a different understanding about how Foreign Languages and Literatures should be conceptualized within an engineering and science environment. I considered MISTI to be a major opportunity to combine the teaching of foreign languages and cultures with the experiential cultural learning that a well designed internship offered, yet my colleagues thought otherwise. And also only in hindsight, I realized how much I underestimated their fear that my approach could lead our department downwards into the role of a "service department." Already on the periphery of MIT's academic culture, my senior colleagues were afraid that the rest of the Institute could see Foreign Languages and Literatures as a kind of in-house Berlitz school without much academic and scholarly standing. I did not share their view, but I admit that such status anxiety is widespread and that it is one of the reasons why foreign language departments are often so hesitant to enter into projects of international engineering education. In addition, those departments are increasingly characterized by a clear division between a large group of lecturers and part-time teachers who only teach language courses and have little professional status and a small group of tenure-track professors who only offer literature and culture courses. Since projects in international engineering education involve colleagues from such different cultures as engineering and the humanities, one lesson learned is how crucial the leadership of deans and department chairs across different schools is. Their leadership and their actions are essential to build trust and mutual respect, and only they can give the reassurance to foreign language departments that they have much to gain from cooperations with engineers and scientists.

BUILDING THE MIT INTERNATIONAL SCIENCE AND TECHNOLOGY INITIATIVE (MISTI)

While continuing to occasionally teach in Foreign Languages and Literatures, I joined MISTI in 2001 to serve as associate director. The experience of moving from an academic to an administrative position is a widely discussed topic amongst academics that I cannot approach here in further details. Since being a scholar is more than performing certain job duties, joining the administrative ranks does indeed pose significant questions regarding one's professional identity. For me, my years with the MIT International Science and Technology Initiative proved to be deeply satisfying. One of the most challenging and fulfilling aspects of my work at MISTI was the constant interaction with a great variety of constituents. I learned to communicate across cultures by mediating between cultures: corporate cultures in which we placed our students and academic cultures across the Institute that we involved in preparing our students. And I learned to appreciate the work of administrators who were able to funnel raw entrepreneurial spirit into a set of policies and procedures that could pass any audit and outside review. As Director of the MIT-Germany Program, I continued to be directly involved with students. Having them in one of my classes and then visiting them in the summer to hear about an exciting research project they were working on side-by-side with young Germans or listening to them talk about their daily encounters with German everyday culture gave me a sense that I was making a difference in their lives.

As Associate Director of MISTI, I was directly involved in building international programs in several other countries besides Germany. Eventually, about 300 students annually were working in all-expense paid internships in our programs in China, Japan, India, Germany, France, Italy, Spain, and Mexico. MISTI is open to all MIT students, about 65% of participants pursue degrees in engineering, about 20% in science, and the remaining 15% come from business, architecture, or the social sciences. About 70% of MISTI students are undergraduates. The large majority of students go for an internship during the summer, but about 20% stay for a semester or start the internship immediately following their undergraduate degree. Each MISTI country program has a faculty member who serves as a director and a coordinator who administers the program. Since several senior faculty members of MISTI have been involved in their teaching and research in the field of international relations, the program is administratively housed in MIT's Center for International Studies.

MISTI works regularly with over one hundred companies and research labs worldwide on two different levels of engagement. While all companies offer tailored internship positions, about thirty of them are members of the corporate consortium and financially support MISTI. Since a large part of MISTI's funding comes from its corporate sponsors, my fundraising activities taught me a great deal about business and corporate research around the world. Over the years, the relationship between MISTI country programs and their corporate sponsors has been extending far beyond sending and hosting MIT interns. The MIT-France Programs organizes successful executive education programs that expose senior level executives of a French oil company to cutting edge research on energy. The MIT-Japan Program has been helping its sponsors for many years in focused recruitment of engineering talent with fluency in Japanese. The MIT-Germany Program regularly brings MIT graduate students and young company employees together for intensive workshops around themes such as "Nanotechnology in the Chemical Industry" or "Car Technology for the Silver Generation." The MIT-Italy Program created an especially close link between MIT and the Polytechnico di Milano through a special seed fund that allows research groups from both universities to work together.

Nowadays, MISTI has become the cornerstone program for international education at MIT. Underlying its success are, I think, two major factors. First, the mission of MISTI was shared by a critical group of entrepreneurial faculty members and administrators who understood the demand from students to gain experience in an international research and work environment and realized the desire of companies to host international talent, certainly for recruitment, but also as an effort to internationalize their labs and work places. Since MISTI has been well funded, our staff could travel regularly and come back with new ideas for projects that they had developed together with our corporate partners.

Secondly, MISTI is a program tailored specifically for MIT. When we started MISTI, it became clear to us that traditional study-abroad would not work in this specific environment. Departments remained hesitant to give course transfer credit to students who had taken courses abroad. While faculty encouraged internships during the summer, many of them openly dissuaded students from being away from the campus during the semester. And we realized that for most students, it was more valuable to have an international internship on their resume than a study-abroad experience in a foreign university. They seek out MISTI because they can learn in an internship something MIT or another university cannot offer them: well-planned and structured practical experience in a real-life work environment.

ENGINEERING EDUCATION AND THE HUMANITIES

During my years with MISTI, I developed a trusted network of colleagues around the topic of international engineering education, and some of those colleagues are also contributing to this volume. Much can be said about international education for engineers. I have chosen three topics that appear to me of particular significance.

WHY INTERNATIONAL ENGINEERING EDUCATION?

This is certainly the widest ranging and most philosophical question in our context. Two strains of arguments should justify and support our work. The first one easily persuades companies, career centers, quite a few of our engineering deans and, of course, students and their parents. Engineers are in the midst of the dynamics of globalization; they often work across continents with colleagues from different companies and increasingly they have assignments overseas throughout their careers. The argument is a powerful one that we need to prepare engineering students for this reality that many of them will face in their future work.

The second type of argument touches upon the ongoing discussion of what it means nowadays to be an educated college graduate. Or more specifically, what does it mean to be a young "educated engineer?" Is the value of a certain annual salary that a professional degree promises in an increasingly globalizing world the only yardstick for a "successful education?"

Undergraduate science and engineering education at MIT, to the complete surprise of many foreign visitors from other technical universities around the world, is deeply intertwined with the idea of a comprehensive college education. All students, regardless of whether they want to become astrophysicists, plan to split genes in one of the life science companies in the Cambridge biotech cluster, or dream of becoming hedge fund managers, need to take a quarter of their classes in the humanities, arts, and social sciences. MIT is currently revisiting the structure of its undergraduate curriculum, yet there is broad consensus among the entire faculty about the crucial importance of the eight-course Humanities, Arts, and Social Science (HASS) requirement. One of the guiding educational principles of MIT reads: "MIT has a unified faculty that takes corporate responsibility for the general education and welfare of our students. As a consequence, of its own intellectual diversity, the faculty encourages MIT students to develop diverse perspectives. These perspectives prepare them to combine multiple modes of inquiry to address fundamental problems." And another principle points out: "Although science and engineering are fundamental components of our culture, their combination with the social sciences, the humanities, and the arts form the core of modern higher education." [5]

The most crucial part in those statements is that an educated engineer has to be able to "combine multiple modes of inquiry to address fundamental problems." And nowhere else, I argue, is this ability more demanded than in what we call "international" or "global" engineering education. Yet the hardest part of accomplishing this goal is to move beyond the mere presentation of different modes of inquiry, and, as the statement calls for, "to combine" them as they emerge out of different disciplines. I feel we still have a long way to go in this respect. It will take academic entrepreneurs to develop such new educational opportunities in spite of an entrenched academic market in which the forces of disciplines rule and regulate. Neither MIT nor most other science and engineering oriented universities have focused sufficiently on the development of such sophisticated interdisciplinary courses that help our engineering students understand the complexities of globalization beyond one discipline.

My own modest attempt in this direction began in 2003 when I started together with some other faculty colleagues at MIT to develop an interdisciplinary "Minor in Applied International Studies," which was successfully implemented in 2005. The core idea was to imbed the students' international experience into a larger curricular structure consisting of courses from the humanities and social sciences. In three courses of the minor, students focus on a particular country or region, and three other courses cover overarching topics such as "Communicating Across Cultures," "Working in the Global Economy," or "Globalization." The strongest reservations about this minor came from my colleagues in Foreign Languages and Literatures who feared that it would gloss over the specificity of cultures. Some saw this minor as a direct attack on more traditional minors in French, German, or Chinese. For those who have worked on any truly interdisciplinary project, this line of argument is not new. Disciplinary depth and expertise is contrasted with a supposedly superficial knowledge of different subjects. Yet globalization forces us to undertake new combinations that differ from established modes of inquiry in one discipline. The minor is now in its fourth year and quickly became the second most popular minor in the School of Humanities, Arts, and Social Sciences at MIT.

As much as I am proud of its success, I would not claim that the minor actually "combines" forms of inquiry. It puts a number of courses in some order, and the final "combination" hopefully happens through the intellectual engagement of the students themselves. It crosses boundaries between Humanities and Social Sciences, but up to this point, it has not brought engineering and science into its curricular structure.

---

[5] Task Force on the Undergraduate Educational Common, *Working Assumptions about MIT Education,* 2004.

EXPERIENTIAL LEARNING

In many of our universities, global learning for students takes place through international internships. Going abroad is a twenty-four hour immersion, a rich and sometimes confusing learning experience. MISTI students work side-by-side with engineers or scientists in companies and labs around the world. On top of learning about another culture through their daily work, students are often confronted for the first time with the intricacies of project management and get a sense of the many different, often managerial roles engineers must fulfill. These international internships are part of a much larger movement within engineering education. Experiential learning through product design classes, project oriented learning, and other hands-on experiences are seen as increasingly important parts of a cutting-edge engineering education. Experiential learning, in a sense, starts from the messiness of a real-life situation. By definition, it forces students to engage in "different modes of inquiry," and, in the best outcome, it allows them to understand the particular contribution of each inquiry to the solution of a problem.

I believe that international education for engineers can play a unique and innovative role within this larger educational movement of engineering education. I am aware how difficult it is to arrange projects across continents, to simulate via role-play telephone conferences about project design, and to actually travel with students internationally. But the results can be deeply satisfying, as those who have ventured out and organized such projects attest.

As an important side note to this, at MIT a disproportionately large number of young women participate in MISTI, confirming overall trends that more young women than men go abroad and engage in international activities. At the same time as I am troubled by the comparative reluctance of young men to go abroad, I am heartened by the fact that global engineering education can be a powerful incentive for young women to enter engineering.

FOREIGN LANGUAGE COMPETENCE IN INTERNATIONAL EDUCATION FOR ENGINEERS

My journey into international education for engineers began as a young assistant professor of German Studies. Throughout this journey one of my goals was to strengthen my own department through increased enrollment, especially in those languages that have seen significant declines in recent years such as French and German. I have already reported my slightly puzzled encounter on the Siemens senior management floor where no German was spoken, even though I considered Siemens a "German" company. The *lingua franca* of engineering and science, of business, and of research is English. Almost 90% of scholarly articles in engineering and science are now published in English. When the German government recently undertook the *Exzellenzinitiative,* an unprecedented competition to found a premier league of German universities by awarding those institutions considerable additional funds, the entire application and selection process was handled in English because experts from around the world were invited to the peer-reviewed process.

Some of my students coming back from their internships in Germany complained that my insistence on their learning German "had been a waste of time" because "everybody they met spoke

English" and their attempt to speak German was aborted by eager German colleagues who wanted to show off their excellent English language skills. So the question is: why should students who we train to work in R&D labs around the world and who often work in multi-national teams in which English is the medium of communication, make the substantial investment in learning a foreign language? Again, I would like to offer two strains of arguments, one pragmatic and one for those of us who believe that speaking a foreign language is indeed part of being an educated person.

Some years ago, I taught a freshman seminar on "Globalization" to a small group of students. As part of the course, we visited Osram Sylvania near Boston. The venerable American company Sylvania (yes, those light bulbs) had been recently acquired by Osram, a part of Siemens. My intent was to bring my students together with middle managers and engineers to learn from them how their daily work had actually changed after their company had been bought by a German multi-national company. Among other things, the language issue came up. "We are so happy that Siemens' company language is English," they told us. "At our age, it would have been difficult to learn German. Now we travel every few months to Munich and have all meetings with our German colleagues in English," they reported. Then the language issue was not really an "issue," I concluded. "Well," one of the engineers responded, "in a way, that's true. All our official communication is in English. But afterwards, in the bar, or in the hallway and lobby during the meetings, our German colleagues sometimes switch back to German, and we don't have a clue what they are talking about. And we know all too well how important those conversations are. That's the reason our boss actually learned German." Consequently, I tell my students that the work in a lab may not require a language other than English, yet for those who feel the need to dig deeper and need nuanced cultural information, knowing a foreign language is essential.

I have to admit that the foreign language teaching profession does not do a very good job of explaining to someone who does not plan to major in a foreign language why it may be a good idea to invest time and considerable mental energy in language learning. While learning a foreign language certainly offers the hope of a much deeper and more nuanced conversation in the host country, the argument that only the mastery of another language will allow someone to gain access to a foreign culture is simply flawed. It is rooted in nineteenth century German romanticism that conflates language with national cultural identity and builds its concept of the nation state around this notion. Those theories did not anticipate that today, millions of educated young people, whether in Beijing, Berlin, or Barcelona explain their culture and themselves in English to the inquiring young foreigner. In addition, simply mastering the linguistic structure of a foreign language by no means ensures successful communication in another country. Communication is so much more than linguistics, as all of those who communicate across cultures realize. And third, as I have pointed out, the relationship between a country's every-day culture and its canon of high culture (books, art, films, music, etc.) is so tenuous, complex, and often contradictory that it should remain mainly the topic of scholarly investigations. Goethe and Flaubert should be read as great writers, not because their works contain some essential "Germaness" or "Frenchness."

Having made those cautionary comments, there are powerful and convincing fundamental arguments why our students, regardless of whether they will become engineers, lawyers, or English professors should engage themselves in learning a foreign language. Let us imagine for a moment a world in which only one language is being spoken. This might be an Orwellian scenario of ultimate efficiency; it would be also a world devoid of what makes us human. Cultural and linguistic diversity underlies the very notion of who we are as human beings. Our humanity expresses itself in such diversity. We sense this when we are in a culture whose language we do not speak and feel the urge to learn at least a few phrases, the most fundamental of which are to say "please," "excuse me," and "thank you." All further language learning extends from there. Paying respect to the diversity of languages and cultures needs to be as much an ethical goal in educating our students as imbedding in their view of the world certain fundamental human rights. We best show our common humanity by acknowledging our diversity and the *lingua franca* of English is neither designed for nor ultimately capable of doing this.

I learned about the term "monoculture" first as a student in geography, long before the heated cultural discussion about "multiculturalism." In agriculture, "monoculture" of crops promised much larger and more efficient harvests—before one plant disease killed the entire crop. It took ecologists such as E.O. Wilson to make us aware of how crucially the survival of our natural environment depends on a wide diversity of plants and animals. Equally, the richness of languages and cultures is crucial for the survival and flourishing of the human race.

I would like to counter the argument about the "usefulness" of learning another language with another, again admittedly philosophical argument. There are certain facets of our existence—actually the most important and most "invaluable" parts of our lives—that defy, even contradict, the reign of utility and usefulness. Friendship, love, art, and religion are among them. To measure those basic human desires with the yardstick of "usefulness" would violate the very nature of those concepts and their experience. I would argue the same is true for having the privilege of learning a foreign language. We can go through life without friendships, but they make our lives immensely richer, we can communicate around the world in English, but being able to speak a foreign language – however imperfectly—offers us a path to discover our common humanity by respecting our cultural difference.

ACROSS THE RIVER CHARLES

In August 2007, I left MIT to become Director of the Office of International Programs at Boston College. I felt that my years as Associate Director of MISTI had prepared me well to take on a larger managerial and administrative role in the field of international education. Such a position simply did not exist at MIT, and managing the international experience of 1,200 BC students in thirty countries has been truly exhilarating. It feels at times like running a small college – from making sure our programs are academically sound to interacting with our partner universities and on-site coordinators to communicating with our students and faculty. At Boston College, almost 40% of a given class will go abroad, mostly into study abroad programs. The Office of International

Programs manages sixty BC programs around the world, mostly as exchange programs with partner universities. We host about 160 international exchange students from those partners on our campus annually.

Internships and research abroad are areas I would like to strengthen in my new position. Another important way that students can interact with a host culture is through service learning projects. At MIT, the "D-lab" project initiated by Amy Smith that focuses on creating technologies for underserved communities showed me what an important role service learning needs to play in international education – not just for engineers. A generous gift from the McGillycuddy-Logue family to Boston College will allow me to conceptualize and put into practice such new ways of international learning. One component that especially interests me is some form of sophisticated reflection after our students come back from their international experiences.

I do miss, though, my interaction with engineers. Boston College does not have an engineering school, even though many students graduate in the sciences. While I have had my doubts at MIT whether undergraduates have enough opportunities to acquire philosophical and humanistic "literacy," I wonder now in such a different academic environment what role "technological literacy" plays in the education of my current students at Boston College. I believe that a liberal arts curriculum such as Boston College offers it within the framework and tradition of Jesuit education should offer its students a stronger exposure to technology— including all the exciting perspectives it offers and ethical challenges and complexities it entails. So there is another bridge to build, and maybe there should be a "study abroad" program between MIT and Boston College.

REFERENCES

Task Force on the Undergraduate Educational Common. *Working Assumptions about MIT Education.* http://web.mit.edu/committees/edcommons/news/workingassumptions.pdf, Massachusetts Institute of Technology, 2004. 263

Widdig, Bernd. *Culture and Inflation in Weimar Germany.* Berkeley: University of California Press, 2001. 257, 259

Widdig, Bernd. *Männerbünde und Massen. Zur Krise männlicher Identität in der Literatur der Moderne.* Opladen: Westdeutscher Verlag, 1992. 259

CHAPTER 13

Linking Language Proficiency and the Professions

Michael Nugent

INTRODUCTION

As I write this[1], I have just entered my third year as the Director of The Language Flagship, a government grant program whose tagline is, "Creating Global Professionals." Having worked over the past ten years on various federal programs to fund international education, I have become increasingly convinced that one of the biggest challenges in maintaining effective programs is establishing a balance between supporting quality innovations and scalable impact. Over the years, I have seen how many of the most interesting innovations started by individual change agents seem to draw national interest but are not easily adopted by additional institutions. Simply put, how can we draw from the lessons we learn from these individual actors and spread them to additional institutions?

As I attempt to show below, when we speak of "globalization" of U.S. engineering education, what we count as quality and effective approaches depends on the viewpoint of the actor. It is my personal contention that one of the most fundamental approaches to creating global education is providing undergraduate students with opportunities to work, negotiate, define, and solve complex problems comfortably in languages other their own. I state this up front because it has been my experience that some involved in global education see language learning for U.S. undergraduate students as a barrier that places unrealistic expectations on faculty and students. This is, however, more of a U.S. viewpoint than a global one. No non-English speaking country would consider a definition of a global professional absent second language proficiency. When it comes to program development and program support, whether on a national or institutional basis, decisions are made by all stakeholders involved based on normative assumptions that shape our personal world views. I hope to convey my own personal trajectory of transformative experiences that helped me shape my views towards global education.

---

[1] This is by definition a personal geography. The viewpoints I express are my own and not those of the U.S. Department of Education or U.S. Department of Defense.

BITTEN BY THE BUG

Likely, the most transformative experience of my life occurred during my eighth-grade year of schooling in Bellingham, Washington when my best friend's mother, a daughter of German immigrants, invited my sister and me to accompany her on a trip to Germany. She told my parents, who were supporting four children on a college professor's salary, that all we had to pay for was the airfare. Though I would have to take off three weeks from school, my mother saw this as a once-in-a-lifetime opportunity, and we cashed in all the savings bonds my grandfather had given us for college to pay the airfare. Though I was only thirteen years old, I still remember almost every single day of the trip, from arriving in the Netherlands, to a traditional Christmas in Southern Germany, to our excursion by night train to Italy. What struck me most, however, were the different languages I heard spoken every day.

As a direct result of this experience, I enrolled in German language classes in high school. Unlike the feeling I had in many of my other high school classes in the mid-1970s, I had a goal: to be able to use the language someday like my European counterparts. At the same time, I also had a dream of and passion for being an accomplished jazz bass player. It was my passion for jazz that ultimately taught me that self-disciplined practice, frequent performance, and ensemble experience paid off in increased skill and ability. This, however, was not the culture of my high school language class, and I decided I would have to put off my dream of mastering a language until I was able to go back overseas. Eager to pursue my interest in jazz music, I skipped most of my senior year of high school and matriculated early at Western Washington University, which offered one of the few jazz studies majors in the country at the time. Here I thrived in a culture of dedicated, goal-oriented musicians.

When I graduated high school, I had almost a full year of university study under my belt and I decided to take the winter quarter to travel to Europe to visit two exchange students, one a Swede and another a German. I had met them while touring Mexico with the high school jazz choir. During the stay at my friend's house in Stockholm, I realized that English conversation on my behalf quickly reverted back to Swedish among friends and family, leaving me feeling as if I had suddenly been isolated in bubble wrap. My few years of German in high school and college, by contrast, provided me a foundation in German, albeit rudimentary, since speaking and listening skills had not been emphasized. Being forced to use it to communicate in my home stay environment in Southern Germany made all the difference in my German skills. It was in this environment where I began to see the multiple levels of complexities involved in mastering a language. Successfully completing homework and classroom exercises was one thing, negotiating on a variety of levels in a home environment was something entirely altogether different. I decided on this trip that one of my major educational goals in life was to be able to engage effectively in discussions on contemporary issues in a second language.

Upon returning to the United States, I decided to continue with foreign language courses while resuming my music studies. After saving up enough money by playing in a nightclub cover band (my first of many professional compromises), I returned to Europe two years later to study

German at the *Goethe Institut* in Lüneburg, Germany. Intent on not being completely left in the dark when visiting my friend in Sweden again, I also enrolled in an intensive summer course in Swedish alongside a conversation course in German at the University of Washington. What struck me about the Swedish course in contrast with the German one was that most of the students learning Swedish were mature professionals who were engaged in research or other various professional exchange activities in areas such as the health sciences. It was clear that most in the class had similar goals and motivations – to learn Swedish well enough to be able to understand the cultural aspects and not be left out in the cold during their time in Sweden. The instruction mirrored this expectation, in stark contrast to the traditional approaches to language instruction I experienced at the same time in German. More importantly, I noticed for the first time the difference a good professor of language could make, as well as the difference it made to be in a class of professionally oriented students who had clear goals and expectations.

When I visited my friend in Sweden, I noticed that the intensive Swedish course at least allowed me to follow the gist of conversations and discussion at dinner tables, bars, and cafes. More importantly, the course provided me a much better window to the Swedish culture. As a result, instead of feeling isolated in bubble wrap at dinnertime conversations, I felt I was sitting at the edge of a large pool, trying to drink every drop I could handle.

My experience at the *Goethe Institut* was also transformative in another way. It is there I learned first hand that, when provided the proper resources, highly trained professionals, and high expectations, an organization could effectively teach languages with results. Not only did the *Goethe Institut* stress all aspects of language learning, it also emphasized historical and contemporary German culture. What also became clear to me through my interaction with other students from institutions across the United States was that my own dissatisfaction with university language instruction at home was not isolated. Twenty-five years later, the Modern Language Association (MLA) Ad Hoc Committee on Foreign Languages would highlight the shortcomings of language education in the United States in an unprecedented report on college and university based language instruction.[2] Essentially, the MLA criticized "the standard configuration of university foreign language curricula, in which two- or three-year language sequence feeds into a set of core courses primarily focused on canonical literature" to be a narrow approach to language teaching.[3] In essence, the report highlighted a frustration with an "organization of literary study in a way that monopolizes the upperdivision curriculum, devalues the early years of language learning, and impedes the development of a unified language-and-content curriculum across the four-year college or university sequence."[4] Of course as a student back then, I knew little about the reasons for lack of interest in foreign language instruction on the part of the university faculty and the institution. Nevertheless, this lack of interest resulted in a lack of expectations they had for students to strive towards any degree of proficiency in languages. In most cases, language classes were structured around a textbook, and the expectations for the courses were to get through the textbook. The contrast with the teaching approach at the *Goethe Institut* was

---

[2]MLA Ad Hoc Committee on Foreign Languages, "Foreign Languages and Higher Education," 2007.
[3]MLA Ad Hoc Committee on Foreign Languages, "Foreign Languages and Higher Education," 2007, 2.
[4]MLA Ad Hoc Committee on Foreign Languages, "Foreign Languages and Higher Education," 2007, 3.

stark. However, as I would learn later, such attention to and skill at teaching language and cultural proficiency did not, in turn, garner respect within the language departments.

By the time I left Germany six months later, I felt very confident in my German, having gained the ability to speak extemporaneously and with relative ease on a wide range of current political and social issues. I was not satisfied, however. My experience at the *Goethe Institut* demonstrated there were new possibilities for language proficiency that I had never imagined. Essentially, they had set expectations for language learning and goals for what one could do if one in fact applied oneself. I set my new goal to enroll as a student at a German university, which would essentially force me to compete with German students in their language. With this new goal, I realized I had to make a choice between my two passions: music and languages. I decided to finish up a degree in German languages and literature, despite the fact that the German Department refused to count any of my coursework at the *Goethe Institut* (despite its reputation and quality, the *Goethe Institut* was not considered an "academic" institution). Nevertheless, because of my higher level of proficiency, I was fortunate to get more one-on-one mentoring in advanced literature study from at least one of the professors who recognized both proficiency and motivation.

My time in Europe had one other impact on me: I decided that in order for me to gain a better understanding of Europe, I needed to learn French. The following summer, I enrolled in an intensive French course, taught by a professor, the late Robert Balas, about whom I had heard much praise for his innovative approaches to language instruction. I was not disappointed. The course was inspiring, and I realized that there was an art to language instruction. Here was a full professor of French literature, who in mid -career had taken the risk to completely change the way he taught first-year French. Not only did he apply at the time the newest teaching methods and theory directly into the classroom, he also began integrating into the teaching process the latest developments in personal computing, software, and digital video. The early 1980s were times that we all look back on now as the Wild West of computer assisted language learning. With the advent of the personal computer, no standard had yet established itself as a common technological platform. Most importantly, he engaged a small group of us who were interested in his small campus "movement" to change the way languages were taught. As a result, we found ourselves spending many hours experimenting in the different ways technology could be used to enhance language teaching.

During this time, I learned another important lesson that would help me understand the work that I am engaged in today. To my amazement, his efforts to improve language teaching were roundly rejected by the other foreign language faculty. As a result, during the early years, Bob Balas invested much of his own money and time into the development of what would later become a state-of-the-art multimedia language learning facility. As Bob always told us, it was not just about technology, it was about rethinking how one used precious time to train different skills using different formats. Due to my engagement and interest, I was also asked to be a peer mentor for exchange students from Japan who were learning English and found that I enjoyed very much teaching languages. Working with Bob Balas was my first experience of many with a change agent who had passion, something I would recognize later as the hallmark of an innovator while working at the Fund for the Improvement of Postsecondary Education.

MEANINGFUL TIME ABROAD

As soon as I graduated, I followed up on my goal to return to Europe and enroll at a German university. I did this against the advice of some of my faculty mentors who had advised me to avoid distractions and proceed directly to graduate school in German language and literature. I moved to Hamburg and enrolled at the University of Hamburg as a student of *Germanistik*, the equivalent at least in name to German language and literature. The decision turned out to be an excellent one in many ways. Since at the time Germany technically had no clear concept of "undergraduate study," students who completed study in this field ended up with a degree far more equivalent to a Masters degree. As a result, I took courses with students who were much older than undergraduate students in the United States. I thrived in the culture of German student life, which at the time was still governed by a level of student academic freedom that was unthinkable at an American university. I also learned the hard lesson that, once I had reached a high level of language proficiency, I would be held to the same expectations and standards of my German counterparts. I enrolled in French and Swedish courses and quickly found that, for all the strengths German universities had, language instruction was not one of them. Whereas the MLA report criticized American universities for a disconnect between the language instruction and the literature faculty, in Germany, foreign language instruction was considered even less part of the German academic tradition and therefore only existed on the fringe.[5]

To help pay my expenses, I took a position at a commercial English school and during the evenings taught English courses to German professionals who had a wide range of abilities and competencies. Given the demand for English instruction in Germany at the time, I quickly found plenty of work. Here I had the opportunity to apply some of the teaching methods I had experienced at the *Goethe Institut* and read about and discussed after hours with my mentor Bob Balas while I was finishing my undergraduate studies. The result led to success and increased demand for my courses. Also, the professionals and business people in my evening classes provided me a completely different perspective on German society from that which I gained through my contact with students at the university. As a teacher of English, I began to see through the rather glib American idea that all Europeans somehow have an innate ability to speak foreign languages. Many of my students were highly successful and busy professionals who were willing to spend their precious evenings and weekends improving their language skills. I especially remember those professionals who came to class with a desire to learn English bordering on desperation. I particularly remember one successful CEO of a mid-sized German engineering company who confided in me over a beer his feelings

---

[5]The clear ideological demarcation between theory and praxis in the German academic tradition impacted many fields of study and has defined much of the discourse over academic reform over the past five decades. I later examined this in my book, Nugent, *The Transformation of the Student Career*, 2005.

of inadequacy because of his relatively weak English skills. Language was clearly something that counted as a core skill to those I was teaching.

During the summer, I used my savings to study French in Tours, France, at the Institute Française de Touraine. There I once again found excellent teaching and a clear purpose to produce higher-level proficiency among students. I also found students from around the world of all backgrounds and professions determined to increase their French proficiency for very specific life and career goals. Speaking for myself, my desire to learn French had moved from the professional to the personal inasmuch as I had fallen in love with a woman in Hamburg whose first language was French. I had met Marie-Françoise years earlier at the *Goethe Institut*, and we had become close friends during my time in Hamburg. She had decided to move to Spain to learn Spanish and encouraged me to come drop down for a visit to see for myself how great life was there. Instead of heading back to my very comfortable life as a student in Germany, I followed my heart and took the night train to Madrid for what I thought would be only a short visit before heading back to my studies in Germany. There I ended up staying for an entire year, having found a job teaching "business English" at a private language school. I enrolled in an intensive yearlong Spanish language program at the University Complutense de Madrid that also encouraged direct enrollment in university courses. By this time, I had become quite experienced in the good, the bad, and the ugly involved in language learning. The courses at the Complutense, though well intentioned, consisted many times of professors who would spend time either correcting grammar exercises or choosing on one of the forty students in the class to read aloud from their favorite passages from classical and contemporary Spanish poetry or literature. One the one hand, I was enjoying the opportunity to learn about an entirely new literary tradition. On the other, I was well aware that I was not being taught language effectively. Nevertheless, I had become fairly self-reliant in my approach to learning languages, establishing my own tutoring exchanges with Spanish students and professionals.

My experience learning Spanish taught me an important lesson about the difficulties of learning a language directly "off the boat," so to speak. Instead of coming to Spain with some foundation in Spanish, I found myself learning language very quickly while at the same time compensating on a daily basis because I was unaware of some of the basics. I always seemed to be functioning verbally way ahead of the very traditional language instruction I was receiving. By the time I left, my Spanish was what one would refer to informally as fluent and highly functional, but in more formal cases, quirky and incorrect. I found that I spoke Spanish often using the same past tense construction as in French, and it has taken me years to try to unlearn this habit. Only later would I discover research supporting the view that students with a solid grounding in the fundamentals of the language perform better when they go overseas – and, this research, incidentally, is one of the fundamental principles in articulating our domestic and overseas programs supported by the work I do today.[6]

---

[6] For discussion on this, see, Davidson, "Study Abroad and Outcomes Measurements," 2007; Brecht et al., "Predictors of Foreign Language Gain During Study Abroad," 1995.

It was in Spain that I realized I needed to make some important decisions. Through connections I had made teaching English to the business community, a large French multinational company offered me a position as director of education in Spain. I was torn between following this career opportunity and returning back to Germany to continue my literature studies. In the end, I decided to seek a different kind of graduate study back in the United States. I had found that my experience in living and working in different European cultures and languages had shifted my interest away from what I thought an American doctoral program in German could offer. Declining enrollments in German in the United States also likely meant a bleak prospective job market in the field. I had realized that in my years living as a student as well as a teacher in Europe, I had developed a very strong interest in policy issues in Europe. Having moved to Spain at the point that it had joined the European Union, I saw firsthand the impact this had on the people in regards to larger social and economic issues, in particular, education and training across Europe. As a teacher, I had spent much time with professionals from many walks of life and I was fascinated by how they got where they were and where they thought they were going. As a result, I had been cultivating an interest in comparative public policy, especially in the arena of professional and academic preparation.

DEVELOPING MY OWN PEDAGOGY

In 1988, I moved to Boston where I was accepted to an innovative summer graduate fellowship program in Second Language Acquisition at Harvard University. A creation of the late Anne Dow, the founder and Director of the Harvard Summer English Language Program, this program attracted on an annual basis the best and most experienced teachers of English as a second language from around the world to work with about twenty new graduate fellows. This program was transformative for all involved. Not only was I trained in the latest language theory and pedagogy by some of the greatest innovators, such as the late Wilga Rivers[7] and others, the summer program also used the latest methods in faculty training and development that teamed seasoned experts in the field with the graduate fellows, all of whom were given a teaching load to teach English to international students who had come from all over the world to learn English for the summer. Many of these master teachers had been fellows years earlier. What Anne Dow had created was a structured environment in which innovators in language teaching could work together, learn from each other, and train a new cadre of language teachers.

In the fall, I was hired part-time by Boston University's Center for English Language and Orientation Programs (CELOP) to teach advanced-level English to foreign students enrolled in degree programs at Boston University. At the same time, I found a job at Roxbury Community College, teaching basic English skills to adults, 50% of whom possessed American high school degrees. The contrast in the students could not have been greater. The next spring, I jumped at the opportunity to join a new program that involved a large-scale consortium of regional universities in the Pacific Northwest. Coordinated by Western Washington University, where I first began

---

[7]Wilga Rivers is widely known as someone who helped bring solid teaching methods to the field of languages. For more information see, Rivers, *Communicating Naturally in a Second Language*, 1983.

my journey in language learning and teaching, this project enrolled approximately 800 Japanese undergraduate students a year in courses taught by U.S. faculty at participating U.S. institutions on topics such as American History and Environmental Sciences.[8] Here I experienced firsthand how partnerships between language and content experts could have tremendous results in creating effective language opportunities for students. However, keeping with my earlier plan to pursue graduate studies that focused on European higher education policy, I had narrowed down my choices for graduate programs with a focus on comparative higher education policy. This research pointed me to Roger Geiger at Pennsylvania State University.

A SHIFT TO HIGHER EDUCATION POLICY

I entered a doctoral program in higher education at Penn State University in 1990, combining European studies and political science with the study of comparative higher education policy. Two things fascinated me in particular. One was the post-war struggle of European governments to accommodate the rapid expansion of their higher education sectors. The other was the developing higher education policy at the European level. This was exemplified most prominently by the European Commission's program to support mobility of students across European member states: the European Community Action Scheme for the Mobility of University Students (ERASMUS).

As a research assistant at the National Center for Teaching, Learning, and Assessment at Penn State University, I coordinated and participated in a number of colloquia that brought researchers from different European countries together to explore common issues of policy reform in higher education. Particularly in Europe, higher education researchers and policy makers had become interested in American-style quality assurance processes, and we had pulled these issues into a series of reports.[9] However, as I quickly learned, such international comparative efforts were difficult and problematic if not undertaken properly. More often than not, researchers were searching for common units of analysis, such as the term "student" or "undergraduate" without clearly examining the underlying differences that these concepts had in their own cultural or national contexts. In many cases, not only was the basic unit of analysis in one country something completely different from another, the research and policy imperatives behind each of these issues were also different.[10] It was these deeper issues that drove my interest in comparative policy, and I was fortunate to receive funding from the DAAD (German Academic Exchange Service) to undertake doctoral research at the University of Cologne where I worked on a comparative analysis of higher education policy and the change in the concept of university study in post-war Germany, Sweden, and the Netherlands.[11] It was here that I realized the advantage of having varying levels of language skills when carrying out comparative policy research in country. Despite the predominance of English in Europe, internal government reports, documents, and records are, of course, not in English. Because of the close

[8]This program is still in operation. Since it was first set up in 1988, over 10,000 Japanese students have studied at the partner institutions in the Pacific Northwest. For more information see http://www.wwu.edu/auap/encontent/about/.

[9]Nugent and Ratcliff, *The International Working Conference on Measurements of Quality in Higher Education*, 1992.

[10]Nugent et al., "Inverse Images," 1993.

[11]Nugent, "The Transformation of the Student Career," 1995.

linguistic proximity of Dutch to German and Swedish, I relied on my primary experience in those languages to explore primary and secondary sources in all three languages.

My work, in essence, turned into a study of the nature of change in the public sphere. Cross-national comparison brought out the underlying idealism and normative assumptions behind the various waves of reform efforts. It also showed how temporal and fleeting ideas could be. By the time a reform was set into motion, society had moved on to other pressing matters. Most significantly, it demonstrated the importance of leadership behind change. When I began my work, I had been warned that no one in the United States cared at all about European higher education policy. However, by the end of the 1990s, everyone with a passport claimed to be an expert on the sweeping European reforms known more popularly as the Bologna Process. What became clear to me was that I really enjoyed examining and working in the area of policy and program development. What especially intrigued me was the potential impact that different types of programs could have on bringing about change. Though much of my research was on European "reform" of higher education, what increasingly interested me was how funding and incentives could help support innovation to bring about opportunities and improvements, especially in the realm of international education. Based on my research interest and my professional and personal experience overseas, I became increasingly interested in working at the macro level on international academic program development.

Upon completion of my doctoral studies, I was able to put theory into practice when I accepted a position in state higher education policy at the newly formed Minnesota State Colleges and Universities. Here I saw firsthand how similar state-level higher education policy initiatives and reform efforts were to those same initiatives in many countries in Europe. I experienced what I observed in many of the reform efforts in Europe—that statewide initiatives and reform efforts would shift or change before they were even put into place, being replaced by new ideas from state coordinating boards or legislatures. Following this work at the state level, I then moved to the national policy scene when I came to Washington to help establish the new Council for Higher Education Accreditation (CHEA), the national representative body for university-based voluntary quality assurance. The late 1990s represented a time that American-style accreditation had become of great interest to European academic organizations and university presidents to help stave off the trend towards increased government intervention.[12]

A MOVE TO INTERNATIONAL PROGRAM MANAGEMENT

My goal of working directly on international higher education program formation was realized when I took a position at the U.S. Department of Education's Fund for the Improvement of Postsecondary Education (FIPSE). Aside from its signature Comprehensive Program, FIPSE had at the time two newly developed international programs: The EU-US Cooperation Program in Higher Education and Vocational Training (the EU-US Program) and the Program for North American Mobility in Higher Education (the North American Program). I was brought on to coordinate the North American Program, which involved the governments of Canada, Mexico, and the United States.

[12]I discussed this later in my book. See, Nugent, *The Transformation of the Student Career,* 2005.

About two weeks after I arrived at FIPSE, we also initiated talks with the Brazilian Ministry about a potential collaboration on a US-Brazil Program. After two years of planning and negotiation, I became the coordinator of the US-Brazil Program in 2000.

The FIPSE International Consortia Programs represented an entirely new approach to federal program management. What made them unique was the multi-country funding structure, whereby FIPSE would fund the institutions in the United States, and the European, Canadian, Mexican, and Brazilian agencies would fund institutions in their respective countries. In each case, a single application was made to the partner agencies, and funding decisions were made collectively by the international funding partners. The European context was especially interesting inasmuch as the European Commission funded institutions in the respective member states of the European Union.

Such a funding arrangement added a new level of complexity to program management, though it also provided an interesting means to create cross-national collaboration and innovation in global-izing the curriculum. The *raison d'être* of the program was to provide an alternative to the traditional study abroad where American students, mainly in the humanities, spent time abroad in special programs that did not necessarily relate directly to their major areas of study. Instead, our intention was to design programs that attracted students in applied or professional fields, such as engineering, who did not normally study overseas. Unlike most study abroad, these initiatives internationalized degree programs, such as Petroleum Engineering, Aerospace Engineering, or even Nursing, by integrating the academic or professional degree programs at institutions here in the United States with those in other countries. In order to achieve this ambitious outcome, faculty at institutions in one country would need to work closely with their colleagues at partner institutions abroad to create curricula and pathways for students to benefit from their time abroad. As this program developed, we began to refer to this experience as "meaningful time abroad." As I discuss below, this effort brought with it a number of challenges at different levels. As one who had a passion for international comparative politics, I was in my element. I saw how these programs provided a rare window into fundamental differences in how academics in different countries conceive and carry out their missions – despite the fact that we often use similar terms.

What I found most compelling about my new position was that I would be working on a project that was a direct offshoot of the ERASMUS program that had intrigued me years earlier. The EU-US Program was first conceived in 1992 by FIPSE's then Director, Charles "Buddy" Karelis, as a pilot effort with the European Commission. Its goal was to provide students from the United States an opportunity to participate in the new ERASMUS program. When the pilot program was started, the idea essentially was to tag on to ERASMUS by providing limited funding to faculty and students to create opportunities for students to enroll directly in participating academic programs overseas.

FIPSE staff quickly found, however, that such a simple goal was rather naïve when dealing with American institutions, faculty, and students. As one of the members of the European Commission responsible for the development of the ERASMUS and the EU-US Program told me when I first joined the team, ERASMUS had been about a lot more than just student mobility. Creating mobility for students meant that European institutions, many of which were either state or federal institutions, needed to become more flexible about curricula, academic credit (which did not exist in many European institutions at the time), and especially about receiving students for shorter terms of study. At FIPSE, creating opportunities for students to enroll directly in their subject areas in many different countries also meant transforming the way faculty and institutions work. We were, in short, attempting to internationalize the curriculum.

GRAPPLING WITH THE CONCEPT OF A GLOBAL CURRICULUM

FIPSE at that time used a flat management style, where each program manager was *primus inter pares*, or first among equals, serving as program coordinator on one program and a staff member on another. All FIPSE staff at that time worked on the FIPSE Comprehensive Program to ensure cross-fertilization of ideas that spanned disciplines and approaches.[13] One advantage I had working on the Comprehensive Program was reviewing on an annual basis an average of 1,600 or more pre-applications that came to our office on almost every topic imaginable.[14] In many ways, the FIPSE Comprehensive Program served as a bellwether to the latest trends in education, whether promising or not. Topics and issues such as outcomes assessment in the 1980s, the advent of the World Wide Web, and "Anytime Anywhere" education in the mid-to-late 1990s came to us as pre-applications in large numbers.

The idea of "global education" was no exception. At one point around 2000 or 2001, we noted a large increase in the numbers of applications that were self-defined as "Global Education" or "Globalizing the Curriculum." Because of the relatively large numbers of applications, we scanned them for a definition in order to break them down into sub-categories. We found that in most cases the term "global education" was not defined, but rather the definition was implied by the activities proposed in the application. The majority of applications seemed to be suggesting an approach that was decidedly not "international education" in its more traditional definition that included any direct exposure to other cultures or study abroad.[15]

Most applicants proposing "global education" made the case of the need to educate students for a global workforce but did not define what specifically they meant by global. The majority of proposed activities usually involved the creation of courses and events that taught students about the world and other cultures. This definition seemed to us at FIPSE to be a reaffirmation of the general purposes of liberal arts education, which in itself is a very noble effort – but not innovative

[13] Much of the long-standing management style and core principles that had so defined FIPSE's effectiveness and success for over three decades has been abandoned since I left. For a history of this unique nature see, Smith, "FIPSE's Early Years," 2002.

[14] Since 1972, FIPSE's signature program has been the Comprehensive Program that funds all areas of innovation in higher education. Projects by topic and discipline can be found in a searchable database at http://www.fipse.aed.org/.

[15] Of course, even the terms "international" and "international learning" are used in different ways. Refer to the FIPSE-funded effort by the American Council on Education "Assessing International Learning" at http://www.fipse.aed.org/grantshow.cfm?grantNumber=P116B040503.

and not necessarily international. Though there were a few grant applications that associated "global education" with the creation of study abroad experiences, few if any focused on language learning. Applications that were self-defined as "international education" tended to propose a study abroad component; however, many did not require language learning. Conversely, few of the applications that focused on improving foreign language learning proposed overseas study as part of the solution.

It is important to point out that of the 1,600 pre-applications to the Comprehensive Program we received on average, we funded only around fifty or sixty projects every year. Because of the competitive nature of the two-stage review process at that time, FIPSE did fund a number of high-profile innovations in international education and global learning. No project better informed the FIPSE team about the possibilities of innovation in globalizing the curriculum than John Grandin's International Engineering Program (IEP) at the University of Rhode Island (URI). This project integrated practices in articulating overseas study, German language and cultural study, subject area study (engineering), and global professional competencies in the workforce. In 1998, I became directly involved in this project when Robert Manteiga, professor of Spanish, received a FIPSE award to begin a Spanish IEP at University of Rhode Island.[16] Working closely with URI, I could see the results of the mature German IEP alongside the new and developing efforts with the Spanish IEP. As is outlined elsewhere in this volume by John Grandin himself, this project essentially was designed to save the German language program at URI from dwindling enrollments. In the end, it transformed an important part of undergraduate education at the University of Rhode Island. In addition, URI decided on its own to create a French IEP without additional federal funding.[17]

As the URI program grew in success, it increasingly became the standard by which we on the international team would measure other projects that proposed to globalize their college curricula. With the help of a small FIPSE grant to seed-fund a dissemination effort, URI successfully began holding annual colloquia that have helped define a movement in international engineering.[18] As a result, the IEP has grown in reputation as the "Rhode Island Model," gaining importance in defining the standard of not only global engineering, but also global education.

LARGE-SCALE CROSS-NATIONAL PROGRAM DEVELOPMENT

Like the Rhode Island project, the FIPSE International Programs were designed to create opportunities for students in fields such as engineering and health sciences who did not normally study overseas. The idea, however, was to do this on a much larger scale. Such an approach put the faculty in the driver's seat of globalizing their curricula. In practice, setting up the FIPSE International Consortia Programs challenged an entire range of practices and policies from all levels, from differ-

[16] For more information about this project see, http://www.fipse.aed.org/grantshow.cfm?grantNumber=P116B980086.
[17] With the help of a Partnership for International Research and Education (PIRE) grant from the National Science Foundation, the IEP expanded to include Chinese in 2005. In 2008, the IEP received funding from The Language Flagship to become a Chinese Language Flagship partner program.
[18] This small investment has turned out to be highly successful in defining the landscape of international engineering. At the time of my writing this piece, they have just completed their 11th Colloquia.

ent bureaucratic traditions of government agencies to equally confusing bureaucratic traditions of institutions.

An additional level of complexity was built into the EU-US Program because institutions were engaging the European Commission rather than the governments of individual countries. In order to bring a European approach to the effort, the Commission insisted in its guidelines that each project have a minimum of three partners, each from a different member state in Europe. Likewise, FIPSE agreed that each project in the United States would involve institutions from three different states. It was difficult enough for U.S. institutions to engage with multiple institutions from different countries in Europe. As we would find out later, an especially difficult added level of complexity came from the fact that U.S. institutions had to learn to work among themselves on a number of administrative and curricular issues. To make matters more difficult, the trilateral North American program had three governments involved in making programmatic decisions. All the countries were using similar guidelines that had been developed in haste during the pilot initiative by taking elements of program guidelines from the Commission's ERASMUS program and pasting them together with elements from the *Education Department General Administrative Regulations* (EDGAR), which were written specifically with U.S. higher education institutions in mind.[19]

At the same time that the EU-US pilot project was developed, the European Commission developed a sister pilot program with Canada. A succession of "offspring" followed, pulling together Canada and Mexico for the North American program and later Brazil for the U.S.-Brazil Program. Working together with our foreign counterparts was not only groundbreaking, it was also highly fulfilling from a personal and professional standpoint. In the U.S.- Brazil Program, for example, FIPSE had been negotiating with CAPES (Coordenação de Aperfeiçoamento de Pessoal de Nível Superior), a part of the Brazilian Ministry, for a few years about establishing a bi-national partnership. Following a summit meeting between then Secretary of Education Richard Riley and the Brazilian Minister of Education, Paulo Renato de Souza, the idea gained heightened importance. The President of CAPES, Abilio Neves, engaged a long-time colleague, Tuiskon Dick, to head up this project on behalf of CAPES. CAPES had a broad mission, which included university research funding and evaluation, higher education statistics, and international exchanges, such as Fulbright. Tuiskon Dick, a former rector pulled out of retirement to help CAPES head up all overseas initiatives, was a formidable collaborator. He immediately saw this plan as an innovative approach to strengthening the relationships between U.S. and Brazilian academic institutions by making Brazil an equal partner and supporting students not only to go to the U.S., but also U.S. students to come to Brazil.[20]

When I visited Brazil in May 2000 to negotiate the terms of the new program, I discovered that President Neves had received his Doctorate in Germany, thanks to the German Academic Exchange Service (DAAD). Tuiskon Dick had also spent many years in Germany as a researcher at

[19]http://www.ed.gov/policy/fund/reg/edgarReg/edgar.html. When I first joined the Department of Education, I kept hearing the question, "What does Edgar say?" I kept wondering who this knowledgeable and powerful guy was!

[20]As a result of the US-Brazil Program, Tuiskon Dick told me at the time that he established similar cooperative agreements with other governments, establishing what he informally called a "FIPSE with Germany" for example.

the Max Planck Institution and oversaw DAAD initiatives in Brazil. As a DAAD Fellow myself, German became our language of negotiation, much to the surprise of our U.S. Embassy translators. A few years later, Michael Hahn, then Cultural Affairs Officer at the U.S. Embassy, told me he believed that the fact we could negotiate without translators had sealed the deal on the project. At the time, all of us on the FIPSE international team spoke at least one other language, which helped build trust and understanding with our partners in Mexico, Europe, and Francophone Canada.

STRUGGLING TO "GLOBALIZE" THE ENGINEERING CURRICULUM

Creation of the FIPSE International Consortia Programs did, in fact, draw early adopters in many different fields. Individual faculty members in engineering were the clear leaders in this area. By 2005, the evidence was clear that annual funding of these programs over the years had created a rather impressive scale of programs. FIPSE and its international partners had funded over 226 separate international projects that involved over 1,439 institutions, 615 in the United States alone. Because some institutions had received multiple grants, we calculated that 894 individual institutions, of which 419 were in the United States, had received funding from the United States and its international partners in Canada, Mexico, Brazil, and the European Union.[21] More importantly, of the 226 projects that FIPSE funded at the time, "engineering and technology" projects made up the second largest major subject area funded across programs, and the 1).[22]

Recently, I contacted six of the early adopters in this program who were involved in engineering-focused programs to gain their perspectives on how these projects may have impacted the study of engineering. All of those with whom I spoke were still engaged in one way or another with their originally funded projects. Many projects dealt with real-world problems and issues in areas such as bio-systems and agro-technology, aerospace engineering, or automotive engineering. Such areas of focus promised students a venue in which they would engage in problem solving in different cultural environments. It was in this context that students saw how culture and social reality affected engineering work, not only in the countries they were visiting, but back home in their own country. As Richard Devon, Professor of Engineering at Penn State University noted, "The most important thing about these programs is seeing your own country in a different way. I tell students that they will not see the U.S. the same way when they return. This is what helps them understand what makes the global economy."[23]

At FIPSE, we could see clearly we had achieved the goal of engaging a relatively large number of U.S. professional and academic degree programs directly with their counterparts at institutions

---

[21] See U.S. Department of Education, "Spotlight on FIPSE," 2005b. http://www.ed.gov/about/offices/list/ope/fipse/spotlight.doc. Many thanks to Don Fischer and Lavona Grow at FIPSE who worked with me at the time to publish this report.

[22] In the case of the US-Brazil program, it was tied for first with environmental sciences.

[23] Telephone interview with Richard Devon, Professor of Engineering Design at Penn State University, on 22 May 2008. Richard Devon's project PRESTIGE: Preparing Engineering Students to Work in the Global Economy was awarded in 2002. http://www.fipse.aed.org/grantshow.cfm?grantNumber=P116J020029.

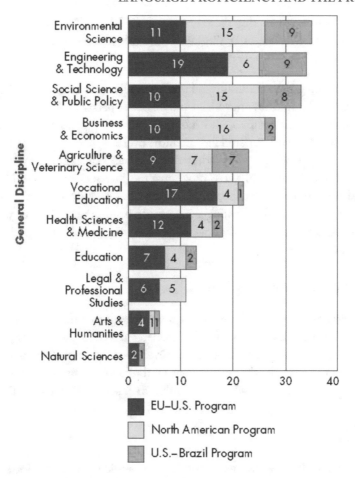

Figure 13.1: FIPSE International Programs: Projects in Each Program by Main Subject Area, 1995–2004.

in other countries. This meant we were finally getting students overseas who came from those fields and disciplines that had long been so poorly represented.

Though these programs were established to impact students, in the end what really mattered was the impact on the academic programs and the faculty. Students came and went, but without a strategy to engage the faculty, these programs would have a short-term impact. Faculty were the glue that held these relationships together, and the level of involvement was very clear from what we saw at each of the annual project directors meetings. With few exceptions, annual meetings were attended by senior faculty, many who, because of the nature of their disciplines, oversaw research labs and projects that had resources that dwarfed the small grants we gave them. Nevertheless, these individuals were committed to making these projects work. Ishwar Puri, now Department Head for Engineering Science and Mechanics at Virginia Tech told me recently that the rewards were both personal and professional:

> It was very enriching both culturally and professionally. As a result of this project I met individuals whom I would have never met and this led to personal long-term relationships with the students, their families, and members in their professional life. I mentored students and worked with our students' mentors overseas, and as a result I developed a confidence in dealing with a wide variety of people, and this helped me communicate with other people. Because of these programs, the intersections of cultures are quite unique, and being forced to intersect across different cultures provided me skills. I don't think I was very comfortable dealing with people from other countries when talking about curriculum, academic institutions, disciplines, etc, but now this project has really had its benefits.[24]

LEARNING ON THE FLY

Because these programs were new, there were a number of issues that were either overlooked or underestimated. In the early years, the issues that challenged us were credit recognition and transfer, student mobility and recruitment, and language. Over time, all governments involved started to see the need to spell out very clearly exactly what kind of support the institution must provide in order to empower project directors, who were finding themselves thwarted by their own academic administrations.

"WHERE CREDIT IS DUE"

No issue vexed us more than dealing with credits. The problem was complicated by a number of factors. First, institutions in the United States, which had over a century of practice with establishing and granting credit, were interfacing with many institutions overseas that did not have a tradition of credit granting. Second, to make matters more difficult, the European Commission, having just created the European Credit Transfer Scheme (ECTS), had provided additional funding to their institutions to encourage them to use ECTS when interfacing with U.S. and Canadian institutions, even though there was no evidence that ECTS would work with U.S. institutions.[25] And lastly, many grantees in the early years had simply overlooked the complexity when they got underway with their programs. A few even tried to set up *ad hoc* credit transfer agreements among institutions. As Ishwar Puri, noted,

---

[24]Telephone Interview with Ishwar Puri, Professor and Department Head of Engineering Science and Mechanics at Virginia Tech, on 20 May 2008. Ishwar joined the EU-US program in 1999 with the project "Globalization and Employability: Benchmarks for Overseas Educational Exchanges." See abstract at http://www.fipse.aed.org/grantshow.cfm?grantNumber=P116J990036.

[25]Despite its name, the European Credit Transfer System was really not originally designed to transfer credit. The system was based more on the recognition of study at another institution than a real transfer of credit. See Teichler, "Recognition," 1990.

The dominant philosophy at the time was that we needed to establish a system for a transfer of credits. However, each of the different components at each institution had to map to each other. As we struggled through this process, we saw that it wasn't about credits, but about processes and outcomes. It was about equivalency mapping. That, for example, A 1 was equivalent to A 2. In the beginning we worked a lot with ECTS and didn't get very far. In the end, we created a system of credit recognition.[26]

The ECTS issue complicated the North American Program as well. Given the widespread perception that "credit transfer" was the holy grail of cross-national collaboration, I was approached by Naomi Collins, former president of NAFSA (Association of International Educators), and Barbara Turlington, of the American Council on Education (ACE), about creating a sort of North American Credit Transfer Scheme, based on ECTS. Though I knew from my own interaction with the projects that students were in fact getting credit, I saw an opportunity for a much more detailed investigation. Given their backgrounds, I asked them to investigate the issue through an evaluation of the North American and the EU-US Programs.[27] During their yearlong investigation, they interviewed a wide range of project directors from both programs.

The result of this evaluation was a report aptly called "Where Credit is Due: Approaches to Course and Credit Recognition Across Boarders in U.S. Higher Education Institutions" that highlighted in great detail the many ways that institutions involved in cross-national programming provide credit for students.[28] In the end, it turned out that the perceived problem of "credit transfer" had overshadowed the creative ways that many of the projects in fields such as engineering were collaborating together. Robert Young, Professor of Engineering at North Carolina State University, remarked that the initial challenge was trying to convince the parents that their kids were not going to waste precious time while studying abroad:

> When I first came to North Carolina State University, we had to reduce the credit hours that students were taking in engineering. The general education requirements also add up. Instead of having a menu of courses, if you target the general education offerings to fit into the engineering curriculum, it offers purpose and certification. Engineering students are very focused on objectives. If you match things with objectives, they will follow.[29]

Though this may seem obvious to the reader today, it is important to keep in mind that few degree-granting programs were engaging in this level of curricular integration at this scale at

[26]Telephone Interview with Ishwar Puri, now Associate Dean of Engineering at Virginia Tech, 20 May 2008. Ishwar joined the EU-US program in while at University of Illinois at Chicago in 1999 with the project "Globalization and Employability: Benchmarks for Overseas Educational Exchanges." See abstract at http://www.fipse.aed.org/grantshow.cfm?grantNumber=P116J990036.

[27]The US-Brazil Program was too new to merit investigation.

[28]See Collins and Turlington, "Where Credit is Due," 2002.

[29]Telephone interview with Robert Young, Professor in Department of Industrial Engineering at North Carolina State University, 23 May 2008. Robert Young received a FIPSE grant in 2002 titled *International Teams in Engineering Industrial Projects: A Cooperative Manufacturing and Production Engineering Program.* http://www.fipse.aed.org/grantshow.cfm?grantNumber=P116M020013.

the time. The FIPSE International Programs put them in the position of having to articulate the program to show how the courses in Brazil, for example, were clearly mapped to those at North Carolina State to get credit preapproved. Essentially, as Young pointed out, they certified Brazilian study in advance to ensure that students would receive credit at North Carolina State University.[30]

What was occurring, as Robert Young pointed out in retrospect, was a change in culture in the engineering programs. At the same time that these programs were developing, important changes were occurring within ABET, then known as the Accreditation Board for Engineering and Technology. Young emphasized that what was a very hard sell in 2001 has become almost a requirement for accreditation today. This change has also been seen in the support from his own institution.

Given the relatively large numbers of institutions involved in the FIPSE International Programs, these early adopters created the beginnings of a critical mass of engagement. In addition, a relatively high amount of crossover occurred in which an institution involved in a EU-US project, for example, would establish a US-Brazil or a North American project. These crossovers became apparent at the annual project directors meetings for each program in which ideas and best practices were shared. On account of sharing best practices, a lot of hard work and perseverance on the part of project directors, and increasing precedence for these arrangements, the credit issue was no longer a stumbling block. This paved the way for increasing numbers of students to get credit for their time overseas.

STUDENT MOBILITY AND RECRUITMENT

Another big challenge was to create a more even distribution of mobility. Some consortia had done an excellent job of achieving mobility and others had achieved very little. The North American program was the most challenging program for mobility. In some cases, no students had gone to Canada, despite the fact that they had first received their funding in 1995. In other cases, administrators at U.S. institutions refused to accommodate students because of what they perceived to be a loss of revenue, despite their having signed on to the terms and conditions of the grant. One of the problems in program design was to figure out how to track student mobility when dealing with different governments and different players. I remember a trilateral North American Program meeting where all three government representatives sat in the front of the room describing three very different perspectives of what the program had achieved in student mobility. Each agency reported different numbers and different results based on their own separate monitoring practices.

Getting involvement was more difficult for some than others, depending on the institution and the project. As Richard Murray, Professor of Engineering at the California Institute of Technology told me, engineering students have a lot of competing options when they are presented with opportunities such as a semester in Brazil. "They were all very interested, but not ready to go. Cal

[30]Interview with Robert Young, Professor in Department of Industrial Engineering at North Carolina State University, 23 May 2008.

Tech undergrads are so goal-directed."[31] Though this was a challenge that faced engineering students across the board, the key to making these programs work was ensuring that students received a high level of faculty direction in their research and studies overseas and credit for their time abroad. Programs that established these pathways efficiently were more effective in recruiting students than those that did not.

In the end, however, students did actually go abroad. By 2005, over 3,500 U.S. students of all disciplines had taken some of their coursework overseas, the majority of whom were in engineering. No one had ever done this before and institutions were not used to accommodating students like these. At the time, most Americans going abroad went through organized study abroad activities; few if any were in engineering or other professional or applied fields. Many of the students who came to study in the United States at the undergraduate level came for the long haul, rather than for a short term, semester-long experience.

LANGUAGE LEARNING

According to Chip Storey, one of the first program officers to work on the FIPSE International Programs, the language issue was largely underestimated when the programs were first established.[32] FIPSE leadership and staff assumed at the time that language would somehow take care of itself; in other words, they expected students to be able to use existing campus-based language resources. The original guidelines did not clearly support language learning. As the programs grew, we found that the language issue manifested itself differently by program. With the exception of the US-Brazil Program, English provided a convenient option for students and projects. Canada was largely English speaking, though some of the best partners were at French-speaking universities. In the EU-US Program, many projects involved institutions in Ireland and England.

The challenge was also related to the fact that the FIPSE programs were supporting short-term student mobility. Based on the original principles of the FIPSE programs, students were supposed to direct-enroll in university courses, which in reality meant that they would have to either come into the program with the requisite language skills or invest a relatively large amount of time to reach the appropriate skill level. To correct the lack of direction on the part of language, FIPSE and its partners strengthened the application guidelines to require applicants to better describe how they would provide opportunities for students to learn language. This clarification, however, did not translate into much additional action on the part of the funded projects.

This lack of attention to language came to a head at the 2000 Annual Project Directors meeting in Copenhagen, Denmark. It became clear to us on the Commission and FIPSE staff that many of the EU-US projects were conveniently ignoring language learning for U.S. students. It did not help that, while we were developing these programs, many of the same institutions we were working with

---

[31]Telephone interview with Richard Murray, Thomas E. and Doris Everhart Professor of Control and Dynamical Systems at California Institute of Technology, 20 May 2008. Richard Murray received a FIPSE grant in 2002 titled *Control and Dynamical Systems Alliance.* http://www.fipse.aed.org/grantshow.cfm?grantNumber=P116M020017.

[32]Interview with Charles "Chip" Storey, 16 May 2008. Chip served as a program officer at FIPSE between 1992 and 1997. He is now a senior policy analyst at the American Institutes for Research.

in Europe were beginning to market special programs in English to attract students from all over the world. We conceded that it was not the Commission's or FIPSE's intention to expect high-level proficiency in a language such as Swedish. Nevertheless, we insisted that if we were going to provide funding to students to go overseas, the project directors could at least provide opportunities for the students to develop basic language skills. It was our position that not providing a required language component cheated the students out of a window into the culture, which was one of the key reasons we were funding these programs. As I had discovered myself years earlier, it was not enough just to visit a country without any background in the language. From the standpoint of the Commission and FIPSE, we were providing support for students to function in a culturally appropriate fashion in other countries, and we considered at least a basic knowledge of language to be the cornerstone of this.

Based on this meeting, the FIPSE's EU Coordinator, Frank Frankfort, and I decided to dedicate additional funding in the form of language stipends for the EU-US and US-Brazil Programs. These were added to the grants in separate line items alongside the mobility stipends provided to the students, and the funds were to be used only for language learning. We provided institutions with basic instructions on how to use these stipends, emphasizing a lot of flexibility in how they use it, including the fact that they could use these funds to establish opportunities overseas.

The response to these language stipends was mixed, clearly reflecting the reality in which the programs were operating. The US-Brazil Program made good use of the stipends, even pooling resources among some of the programs to establish summer intensive programs in Brazil and, in other cases, hiring instructors to teach Portuguese at U.S. institutions. The EU-US Program was particularly complex because of the larger number of less commonly taught languages, such as Swedish, Dutch, and Finnish. This issue became all the more difficult with the addition of countries such as Hungary and the Czech Republic. For programs that worked closely with France, Spain, or Germany, for example, the problem was not as pronounced. For example, Richard Devon told me that he was surprised at the number of engineering students who joined his program with some language experience. Penn State, it turned out, was large enough to draw students who had some French or Spanish.[33]

The pushback on the language issue came from both sides of the Atlantic. Some partners in Europe saw language stipends as an unnecessary add-on to the effort. I remember one directors' meeting in which a professor from Sweden vehemently opposed our efforts to provide opportunities for American students to learn some basic language skills while in the program. In his opinion, learning Swedish "took a lifetime" and no one had the time or the resources to support this. Further, according to him, why should American students learn a language that no one spoke? Though many of his European colleagues countered his opinion, he was stating a sentiment shared by some U.S. and European faculty. Thinking back on my own personal experience of having taken a summer intensive course in Swedish, I felt that the students would lose out on a great opportunity if the faculty completely neglected the language.

---

[33] Richard Devon interview, 22 May 2008.

Not everyone agreed with the sentiments of that Swedish professor. There were clearly two camps of thought about language learning: those who thought it got in the way, and those who believed it was the most essential part of the experience. Some directors were in agreement that decisions such as language should not be left up to students as an option. As Ishwar Puri told me, "Language is a basis for culture. Our students still don't recognize this. It is not a problem for engineering. It's one for American education. This will be a harder issue to resolve."[34]

In the US-Brazil Program, the language issue played itself out quite differently. English was much less spoken in Brazil, even by the well-educated. Spanish offered a very substantial bridging language, thereby making it easy to bring students up to speed. Nevertheless, the new programs had a number of language hurdles to overcome. In the case of North Carolina State University, the study abroad office required two years of language before students could participate in university-approved study abroad. Adding this sort of requirement to an engineering curriculum was out of the question. The solution, therefore, was to create a special "Spanish for Engineering" course to bridge students into the Portuguese language. As Robert Young remembers, "Originally there was resistance from the Foreign Language Department. However, they discovered that engineers make some of the best language learners. Soon language faculty began to fight to teach this course."[35]

This course was followed by an intensive Portuguese course created especially for these students at their partner institution, the Universidad Federal de Rio de Janeiro. This new course has been so successful that it is now used by a group of U.S. institutions involved in the US-Brazil program.

Though language learning remained a challenge, it was Robert Young's opinion that most engineering faculty are not opposed to language learning, "Since so many engineering faculty members are foreign-born themselves, they see the value in language. However, given the choice between having students learn a language or learn differential equations, they will most always choose differential equations."[36]

Young, had a very personal conviction that language was important:

Ultimately, I didn't care about their engineering classes. I cared about their language. I had personal experiences with this with my own children. I think the students follow the faculty and their parents. If the faculty consider it important, the students find it important. I find that there is an attitude change among engineering schools.[37]

It is difficult to generalize about the commitment to language learning from engineering faculty, especially given the differences among the US-Brazil, North American, and EU-US Programs. Many of the attitudes about language were squarely based on pragmatism. Given the relatively small amount of money, short time overseas, lack of language teaching resources, and relative difficulty of some of the European languages, the majority of projects tended to avoid the language issue, despite the available language stipends.

[34] Ishwar Puri interview, 20 May 2008.
[35] Robert Young interview, 23 May 2008.
[36] Robert Young interview, 23 May 2008.
[37] Robert Young interview, 23 May 2008.

Despite the mixed results the FIPSE programs had with getting faculty to support students learning languages, I have seen a clear change in the attitudes of many faculty and staff involved in projects. Many of the challenges such as credit recognition that faced those early adopters seem to have disappeared. Instead, many of the practices have been institutionalized and program curricula reflect cross-national comparative approaches that include in-country experience for students. In schools of engineering, new positions have been created for international program directors in dean's offices. Though I have not formally polled all the former project directors, many FIPSE project directors have found themselves in leadership positions dealing with international education on their campuses. Many I talked to in the past and today felt that without the support from FIPSE, there would not have been the experience or critical mass to bring about these arrangements.

After about eight years at FIPSE, I took another position at the U.S. Department of Education as Chief of the Advanced Training and Research Team, International Education Programs Services (IEPS). IEPS oversaw ten initiatives more commonly known as "Title VI Programs" and four programs under the Fulbright Hays Act that oversee a relatively large investment, in comparison to FIPSE, in long-standing international education programs. Many of these programs were first put into place as part of the National Defense Education Act of 1958. My experience in this position is a story in itself and too involved to cover in this piece. Suffice it to say that during my tenure in this position, I became all the more convinced that the relative location of an agency within a larger organization makes all the difference in trying to bring about effective program management and support. In other words, programs with the best-laid goals and intentions can be hindered by multiple layers of ineffective administrative structure and inadequate attention to the importance of good personnel.[38]

PUTTING LANGUAGE FRONT AND CENTER

In 2007, I took a position at the National Security Education Program (NSEP) as the first Director of the National Language Flagship Program. I had always respected NSEP as a program that rivaled FIPSE in its level of innovation. The mission of The Language Flagship is to globalize education and change the paradigm of language learning by creating opportunities for students of all majors to attain a professional level of foreign language proficiency (technically defined by the Interagency Language Roundtable (ILR) as being "able to speak the language with sufficient structural accuracy and vocabulary to participate effectively in most formal and informal conversations.")[39]

As I look back, I realize that my work with the Language Flagship represents in many ways the culmination of all my personal and professional experiences to date. The passion I developed thirty-five years ago for languages and different cultures is now transferred to a similar passion for providing opportunities for thousands of American undergraduate and graduate students of all majors to learn

[38] For more information about the Title VI and Fulbright Hays programs see, O'Connell and Norwood, *International Education and Foreign Languages,* 2007.

[39] The Language Flagship sets the Interagency Language Roundtable (ILR) Scale of 3 or the American Council on the Teachers of Foreign Language (ACTFL) scale of Superior as its target for all successful students completing The Language Flagship. For more information on the ILR, visit http://www.govtilr.org/ILR_History.htm.

foreign languages. Because the Flagship program provides opportunities for students to commit to learn Chinese, for example, alongside their chosen majors, such as engineering or public policy, students do not have to make a choice between studying language or another discipline. In many ways, students end up with majors in both. In order to achieve this goal, the Flagship program supports colleges and universities to change the way they teach languages and the disciplines. Language learning is key to the Flagship undergraduate experience, ensuring a rethinking of undergraduate education that involves content-based language learning, overseas direct enrollment in the discipline or subject area, and internships overseas. In entering into a program, students of all majors agree that they will strive to master their subject area within this new context. Though all students have a target level of professional proficiency, or Superior as described on the scale established by the American Council for the Teachers of Foreign Languages (ACTFL),[40] a successful graduate has a strong background in the related cultural and historical roots of the languages as well.

Over the years, my colleagues and I have always been daunted by the nexus between program quality and scale. Unlike the larger scale investment in the FIPSE programs mentioned above, The Language Flagship currently consists of about twenty domestic centers and programs as well as nine overseas Centers involving over 100 faculty, administrators, and staff at institutions in the United States and abroad.[41] The funding provides center and program directors the opportunity to work closely together on restructuring their undergraduate programs to create global opportunities for students of all majors. In 2008, the University of Rhode Island International Engineering Program joined The Language Flagship with a Chinese program. URI has also begun to examine its current German, Spanish, and French IEP programs to establish higher language proficiency goals that match those of The Language Flagship. However, they will not do it alone, but rather as a community of individuals attempting to achieve the same thing. Such an approach allows early adopters and innovators to share their successes and failures, their lessons learned, while at the same time working together to establish new administrative practices at large institutions that are, by necessity, bureaucratic.

One thing I have learned directly working with change agents over the years is that even the smallest change and innovation takes a lot of hard work and dedication and almost all of this energy comes directly from the agent. As I had observed almost thirty years earlier as a student, effective innovation is not usually enthusiastically embraced by other members of an organization. In many cases, I have observed that the more effective and innovative the idea, the higher the level of resistance to the change. Though I realize that I am offering an opinion that may seem to be a sweeping generalization, my opinion is a result of funding over 400 separate initiatives at many different types of institutions. Though outside funding, federal or otherwise, is crucial to give ideas leverage and momentum, such funding usually has little effective impact without the change agent. That said, if an initiative is too focused on one individual it runs the risk of diminishing institutionalization and sustainability. Institutionalization and sustainability require longer-term engagement, not just on

---

[40] See ACTFL Proficiency Guidelines Speaking at http://www.actfl.org/files/public/Guidelinesspeak.pdf.
[41] For more information on the The Language Flagship visit www.Thelanguageflagship.org.

the part of the funder, but on the part of the larger community of innovators involved in the effort. In the case of the FIPSE International Programs, much was learned by bringing these innovators together to share their success. However, it is still not clear whether these programs will be able to sustain long-term relationships. For this, we have worked to build a community of agents, or innovators, which helps develop a support system to normalize change over time.

What I really find interesting, however, is talking with the students who have committed themselves to this arduous effort. Students, more than anyone, understand what our effort is all about and do not need to have the idea explained to them. Over the years, I have heard the story of how ambitious students are sometimes discouraged from combining advanced language learning and overseas study with their major. My personal experience of combining disciplinary research with language and cultural study has been so rewarding to me that I am committed more than ever in trying to make this national effort work. Each time I meet with these students, I see the same passion that I had as a young student. It is for this reason that when people ask me how I can be so passionate about what I do, I tell them that I have found a true vocation, or calling, in life.

REFERENCES

Brecht, Richard D., Dan E. Davidson, and Ralph B. Ginsberg. "Predictors of Foreign Language Gain During Study Abroad." In *Second Language Acquisition in a Study Abroad Context*, edited by Barbara F. Freed, 37–66. Philadelphia, PA: John Benjamins Publishing, 1995. 274

Collins, Naomi and Barbara Turlington. "Where Credit is Due: Approaches to Course and Credit Recognition Across Borders in U.S. Higher Education Institutions." Washington DC: American Council on Education, 2002. 285

Davidson, Dan E. "Study Abroad and Outcomes Measurements: The Case of Russian." *Modern Language Journal* 91 (2007): 276–280. DOI: 10.1111/j.1540-4781.2007.00543_13.x 274

MLA Ad Hoc Committee on Foreign Languages. "Foreign Languages and Higher Education: New Structures for a Changed World." New York: Modern Language Association, May 2007. DOI: 10.1632/prof.2007.2007.1.234 271

Nugent, Michael. "The Transformation of the Student Career: Change in the Concept of University Study in Germany, Sweden, and the Netherlands."Ph.D. diss., Pennsylvania State University, 1995. 276

Nugent, Michael. *The Transformation of the Student Career: University Study in Germany, Sweden, and the Netherlands*. New York: RoutledgeFalmer, 2005. 273, 277

Nugent, Michael A., James L. Ratcliff, and Stefanie Schwartz. "Inverse Images: A Cross-National Comparison of Student Persistence in Germany and the United States." In *Das Amerikanische Hochschulsystem: Beiträge zu seinen Vorzügen, Problemen und Entwicklungstendenzen* (Zeitschrift für Hochschuldidaktik 17 no. 2–3), edited by Hans Pechar. Vienna, Austria: Österreichische Gesellschaft für Hochschuldidaktik, 1993. 276

Nugent, Michael and James L. Ratcliff, eds. *The International Working Conference on Measurements of Quality in Higher Education.* University Park: National Center on Postsecondary Teaching, Learning, and Assessment, 1992. 276

O'Connell, Mary Ellen and Janet L. Norwood, eds. *International Education and Foreign Languages: Keys to Securing America's Future: Committee to Review the Title VI and Fulbright-Hays International Education Programs.* Washington DC: National Academies Press, 2007. 290

Rivers, Wilga M. *Communicating Naturally in a Second Language: Theory and Practice in Language Learning.* Cambridge, MA: Cambridge University Press, 1983. 275

Smith, Virginia. "FIPSE's Early Years: Seeking Innovation and Change in Higher Education." *Change Magazine* 34 no. 5 (2002): 10–16. 279

Teichler, Ulrich. "Recognition: A Typological Overview of Recognition Issues Arising in Temporary Study Abroad." Werkstattberichte Wissenschaftlisches Zentrum für Berufs-un Hochschulforschung der Gesamthochschule. Kassel, Germany, 1990. 284

U.S. Department of Education. "Grants Policy and Oversight Staff, Education Department General Administrative Regulations (EDGAR)." Washington, DC: Department of Education, 2005. http://www.ed.gov/policy/fund/reg/edgarReg/edgar.html.

U.S. Department of Education. "Spotlight on FIPSE International Programs." Washington DC: Department of Education, 2005. http://www.ed.gov/about/offices/list/ope/fipse/spotlight.pdf. 282

CHAPTER 14

Language, Life, and Pathways to Global Competency for Engineers (and Everyone Else)

Phil McKnight

Among the many pathways leading to a match between the goals of students, industry, and academics, the International Plan emerged as a multifaceted collaborative effort at Georgia Tech to globalize the educational process. The success of the International Plan led to Georgia Tech being named recipient of the 2010 Andrew Heiskell Award for Innovation in International Education, as well as of the 2007 Senator Paul Simon Award for excellence in practices, structures, philosophies, and policies that internationalize the campus.

Since foreign language departments have often (justifiably) been seen as unwilling participants in collaboration with science and technology units, the first part of this essay will describe the emergence of my own perspective and the anecdotal and intellectual encounters that led to my conviction and desire to engage a far-reaching collaborative effort between my unit and others at Georgia Tech. A description of transitions follows along with an appraisal of difficulties and successes in implementing the International Plan, in which the School of Modern Languages played a key role. The essay finishes with reflections on the impact on students' lives and values beyond newly enhanced skill sets honed during their study/work abroad experiences.

A SECOND LOOK

As US colleges and universities began to emerge in the first decade of this century from the prolonged decline of foreign language study and study abroad,[1] pragmatic approaches to teaching and learning foreign languages have gingerly come forward from the shadows, coinciding with educational changes made to globalize engineering and technology. This is not to say that foreign language departments have moved very far away from the traditional forms of their disciplines, which consist primarily of a curriculum in high culture and literature or of an "areas studies" track that leads to careers generally restricted in scope to teaching language and literature. Most language departments

---

[1]Cf. Hudzik, et al., *The State and Future of Study Abroad in the United States*, 2004; and IIE, "U.S. Study Abroad up 8%," 2008.

have not changed their time-honored approach even though students have long since adjusted theirs to the 21st century. But as Simon Marginson points out, higher education has become "a central driver" of global connectivity and the flow of "people, ideas, knowledge and capital" which has transformed our enterprise into the "Global Research University" with its high level of student and faculty mobility, and "cross-national research and learning."[2] The need for foreign language and intercultural skills within the context of this massive new mobility, which has recently been argued at great lengths, has become more than obvious, and so has the obligation of foreign language departments to expand their offerings to be inclusive of students' needs.

My own experience took me through several levels of fluctuation between an ivory tower isolation immersed in 18th century German literature and society, and real world connections that may have actually materialized in the 1980s during my prolonged stay in East Germany. But truth be told, the 1990s forced a second look at my profession as its enrollments declined with the dissolution of political tensions in Germany that had been critical to American military and political interests.

Nevertheless, an unusual background that included seven years working in a motorcycle shop and a ten-year hiatus from academics due to financial necessities associated with single parenting may have provided some fundamental inclination in my character to seek connections between the typical aesthetic realms of literature and the real world. The idea for me was not to reject humanistic thought, humanities, critical thinking, etc., but to overcome the sense of self-inflicted isolationism that often occurs within the humanities, which in some respects have failed to deliver the very product they purport to represent. I thought there could be a way to improve how the US conducts business and politics abroad by adding communication skills to the disciplines that enable us to be both collaborative and competitive in foreign countries. Digging up a 1984 article I published in conference proceedings for *Applied Language Study* in Stillwater, Oklahoma, I discovered that I had actually used the words, "we are the senior engineers"[3] of our educational discipline, and therefore must share the blame for not having integrated ourselves into the mainstream of the lives of students during and beyond their educational experiences. We had engineered a faulty design and the marketplace was moving on.

To regress even further: as an MA student at the University of Colorado in the late 1960s, when foreign languages were flourishing (thanks to Sputnik!), I had organized a symposium addressing the question, "why do we study Germanics," framing it with Goethe's famous 18th century play *Faust*, in which the protagonist rejects his isolated and frustrating life in a university research lab to answer the call to, "flee out into the wide world" (*Flieh! auf! hinaus ins weite Land!*). If the principle work of the principle "employer" of German professors (at the time probably 25% of research in the field focused on this author and period) admonished us to get rid of this pile of books and test tubes to find out "what the real essence that held the world together" was (*Dass ich erkenne, was die Welt/ Im Innersten zusammenhält*), why were we not taking his advice? Why was their no connection between what we read and learned, and life itself?

---

[2]Marginson, "The Rise of the Global University," 2010.
[3]McKnight. "Foreign Language Study," 1984, 20.

My professors, steeped in theory and the purely academic—a word which also can mean that an issue is no longer relevant—balked at this uncomfortable questioning, avoiding the existential question altogether. Much later in East Germany, where I met a number of authors, I realized that although writers may need scholars to ensure posterity, they were very aware that their main audience was the common reader and how he/she would absorb their writings into their own socialization process, into how they thought and felt about life, politics, and society. In my own case, it took many more years for me to understand that Faust himself was saved in the end from his deal with the devil by becoming a civil engineer and building structures that served the good of humanity, with which he overcame his egotistical self-indulgence. This is still not an interpretation that can be readily found in the annals of German literary theory.

Nor would I necessarily propose that engineering technology is an automatic voucher that justifies all that we build, since some ethical, environmental, and cultural issues must be a part of our endeavors if we are to be global citizens thinking about the deeper impact of what we produce. For example, driving out a complete culture of people from the lowlands of Cambodia into the mountains in order to accomplish massive deforestation in the valleys is not necessarily an engineering or logistics feat to be proud of, especially in view of the new life these people were forced into, namely earning their living by digging up old wartime mines in the mountains at risk of life and limb to sell them as scrap metal.[4] And the implications of the agricultural engineering and industrialization in the rain forests of Brazil are even more far reaching, as tracts of land as large as France and Portugal combined have been deforested to plant soybeans—which never grew naturally in Brazil and therefore require massive amounts of fertilizer and pesticides that run off into the groundwater. Not to mention the disastrous displacement of indigenous cultures and the pollution of their water sources, or the loss of CO^2 retention inherent to the rain forest.[5] Would any of this have been done differently had experts in culture, language, and society been able to exert their influence in a credible manner, or by integrating their views collaboratively with those held by business, engineering, technology, and policy sectors?

A recent position paper by the Modern Language Association has urged departments to rethink their approaches to language, literature, and the structure of their enterprises. The wider context for the changes recommended is not really addressed. I maintain that the changes are partly due to the fact that students who learn foreign languages today and who subsequently engage themselves in an experience abroad now come from a wider range of disciplines than in the past. Statistics show that in 2007-08, some 11% of all students studied abroad. Only 3.1% of engineering students and less than 10% of computer science, math, and physical sciences students (combined total) studied abroad. Most students came from the humanities, social sciences, fine or applied arts, and foreign languages—for which, astoundingly, only 6.2% study abroad.[6]

---

[4] Summary from U.S. Department of Education and the Center for International Studies at the University of Chicago, *Environmental Challenges across Asia*, 2007.

[5] Cf. Wagenhofer, *We Feed the World*, 2005.

[6] IIE, "Open Doors 2009 Report on International Educational Exchange," section on Fields of Study, 2009. In 2007/08, the leading majors of Americans studying abroad were social science (21.5%), business and management (20.2%), humanities (13.3%), fine

But the influence of globalization has produced both a new awareness and a sense of urgency for the current global student generation in engineering and technology with respect to how their career trajectories will be defined in the future. Industry leaders have often indicated a strong desire to remain competitive in a global environment by hiring graduates with language and intercultural skills combined with work abroad experience,[7] and many students now seek the training and education that will make them competitive in their job search and enhance their long-term careers within the so-called "flat" world.

I base this assumption not just on anecdotal interaction with students, but also in part on the doubling of enrollments in foreign languages at Georgia Tech over the past five years, which has taken place with no general foreign language requirement, so that per semester, the percent of Tech students taking foreign languages has grown from just under 10% to 21.7%. The current national average is 8.7%, down from 16% in the 1960s. Well over 60% of these students are engineering and computer science majors, indicating that the spike in enrollment numbers signifies a substantial increase in engineering students seeking to become proficient in another language and culture. Most students and faculty have realized that educational needs in the 21st century reflect the challenges and opportunities of globalization and technological developments in international trade, computing, media, and information exchange, creating much stiffer competition. Graduates prepared to act as intercultural mediators are needed across the board.

Even as engineering units are rapidly internationalizing, most foreign language departments continue to struggle to adapt to students' new aspirations and needs. In fact, many over the years have separated their enterprise into a two-tier system consisting of language teachers and literary/theory scholars. The language group typically is paid much less, usually does not operate in tenure track lines, and with caveats of "business French," etc., has much less prestige than do the theorists, who typically teach literature and theory, using primarily English in the classroom, with readings in the target language. More importantly, this latter group has absolute power over curriculum development, hiring, and tenure decisions.

I recall a meeting of department chairs at a German Studies conference in New Orleans in 2003, during which I was the only one of three speakers presenting ideas for new directions who actually presented in German. Unlike me, a near-native speaker, the majority of the participants were native speakers. Afterwards, the entire discussion took place in English, an indicator of how language departments relate to the very languages they teach.

Reacting considerably later than major government studies on the need for foreign languages,[8] the Modern Language Association concluded that Americans needed "to understand other cultures and language," that the "language major should be structured to produce a specific outcome: educated

---

or applied arts (8.4%), followed by physical sciences (8.4%), foreign languages (6.2%), education (4.1%), health sciences (4.5%), engineering (3.1%), and math or computer science (1.6%).

[7] Gutierrez et al., "The Value of International Education to U.S. Business and Industry Leaders," 2009.

[8] These reports all stressed the critical national need for foreign language skills in all disciplines. See, among others: Gross and Lewis, "Education for Global Leadership," 2006; McPherson et al., "One Million Americans Studying Abroad," 2005; Obst et al., "Meeting America's Global Education Challenge," 2007; and Gutierrez et al., "The Value of International Education to U.S. Business and Industry Leaders," 2009.

speakers who have deep translingual and transcultural competence;" and that curricula should be designed so that students are "trained to reflect on the world and themselves through the lens of another language and culture." The MLA called for an end to the hierarchical approach that "creates a division between the language curriculum and the literature curriculum and between tenure-track literature professors and language instructors in non-tenure-track positions"[9] in order to create opportunities for interdisciplinary activity that includes foreign languages. In spite of this ground-breaking exhortation for change, a close read of the recommendations reveals little pragmatism and quite a bit of same old same old. The suggestion to be more integrative with other disciplines appears almost as an afterthought and lists the standard partners such as international studies, history, anthropology, music, art history, philosophy, psychology, sociology, and linguistics, with law, medicine, and engineering bringing up the rear, omitting business, economics, and computing altogether.

The paradigms for "new" approaches were appropriate and astute: language study can be used to "understand how a particular background reality is reestablished on a daily basis through cultural subsystems such as: the mass media," or "to see literary and artistic works as projection and investigation of a nation's self-understanding," that language is the key to "the social and historical narratives in literary texts, artistic works, the legal system, the political system, the educational system, the economic system, and the social welfare system" and reveals "local instances of major scientific and scholarly paradigms" with "the cultural metaphors these have created, and their relation to the national imagination" and "stereotypes, of both self and others, as they are developed and negotiated through texts, symbols or sites of memory."

But the kinds of "new" courses recommended by the MLA reveal a much less ambitious path to change, courses which would focus "for instance, on a period, an issue, or a literary genre," presenting "an in-depth study of cross-cultural influences." Sample courses include: "the Crusades in the Middle Ages; the Silk Road; literature and opera; the sonnet across four national literatures; turn-of-the-century Vienna, Paris, and London; literature and science; and interconnections between Germany and the United States."[10] I would contrast these suggestions with courses being taught at Georgia Tech, such as Food, Culture and Society, which examines food production, consumption, and commerce as a gateway to the exploration of Hispanic cultures, identity, community, sustainability, and effects of globalization (taught in Spanish); or 3D Role Playing Games, in which students interact with a Second Life platform to learn about Japanese life and culture (taught in Japanese); or Perspectives of German Media, which examines the nature of propaganda, how media represent politics, business, and culture, and media representations of technology and environmental issues (taught in German); or our intensive summer Language for Business and Technology programs taught in nine countries abroad, to name a few examples.

---

[9]Geisler et al., "Foreign Languages and Higher Education," 2007.
[10]Geisler et al., "Foreign Languages and Higher Education," 2007.

TRANSITIONAL GEOGRAPHIES-EMERGING VISIONS

My own change of direction began while still employed in the motorcycle shop trying to make ends meet as a single parent when I enrolled in a summer program at the University of Virginia. The program was designed to transform Ph.D.'s in the humanities and provide them with tools to switch careers. The six weeks included cramming of business, finance, marketing, and IT into classes, plus extensive networking in the Washington DC area. The faculty were outstanding and my apprehensions about "consorting with the enemy" as a dedicated humanist were alleviated greatly by their presence and performance. The afternoon networking sessions were foreign to my nature at the time: calling strangers on the phone had always seemed to be an infringement on others' privacy, and to seek interviews and "advice on careers" did not seem genuine or sincere. But the interviews proved somewhat useful, mainly teaching me not to fear them, and the people I met in DC were genuinely helpful.

We, the humanities Ph.Ds, had some spare time to talk and stimulate each other's ideas. I reflected that I had adapted well to single parenting—tilled a gigantic garden with vegetables that I harvested and canned for the winter, even had a chicken that laid an egg every day for my daughter's breakfast, found time to cruise up the Colorado canyons on a shop-owned used Honda CB 750 or off-road trail biking on an XL 250, and even romp on the reservoir with Kawasaki jet skis—but it was gradually dawning upon me that all this might not lead to a long-term stability and future education for my daughter, then entering the second grade.

Sometimes traveling Germans would come into the shop and we would speak German, reminding me of my years in Bonn, Tübingen, and Berlin, and what I thought might have been. I think that deep inside I did not want to lose the international part of me and the love for Germany. Perhaps, this is what led me to think international as I was considering more pragmatic approaches to life and imagining a new career. Actually, I had never intended to make German a career when I first returned to college after sitting out a year and a half, but some special professors had caused me to reinvent myself, and I was trying now to reinvent myself to think differently about material values, or at least to include them in my thinking. Self-reinvention, as I learned, is a healthy way to prevent stagnation and fuel the creative fires.

So I returned to Colorado with a plan to approach *Colorado Business Magazine* as a free-lance writer to do an article on international trade and free trade zones. When they actually agreed to let me take this on (for a nominal payment as well!), it became my entry ticket into government and executive offices to discuss with them their plans for globalizing their businesses. I joined the International Trade Association of Colorado (ITAC), and I began to do my homework, something I had been very good at in college. I decided to gauge their potential interests in creating a free trade zone in Colorado to reduce inventories and save costs on imported parts, which could remain warehoused for just-in-time delivery, and which therefore would not be subject to import levies until they actually touched the assembly line.

I intended to use this networking opportunity to find myself a job in one of these companies, or better yet, to match my yearning for the international, working in a foreign trade zone that I would help develop. Taking days off from the motorcycle shop to drive up to the Denver-Boulder area, I was, in fact, able to confirm strong interest by major companies like Storage Tek[11] and by the Colorado Economic Development Office, which thought they might be able to find investors. Perhaps a self-styled biker-humanist with a Ph.D. in German could not succeed in this environment, and no doubt I was projecting a much too ambitious entry level for myself, but these events most definitely were opening new pathways and new ideas.

RETURN TO ACADEMIA

However, in the midst of these activities, along with a failed attempt to find an investor in the motorcycle shop—which by then was on the market—who would install me as the new manager, an instance of "divine intervention" took place. In August of that year, the University of Kentucky asked me to take a one-year appointment to replace a tenured professor who had decided to give up all her material possessions to Baghwan Shree Rajneesh and join him in Oregon. I am not sure how this turned out for her, as just a few years later in 1985, owing to his "free-love" policies and the manner with which he would drive many of his some eighty-five Rolls Royce's in a caravan through town, he was deported for violation of US immigration laws.

As for me, I realized that all the years I had dedicated to Germany should not go to waste. I won the tenure-track position the following year and re-immersed myself into 18th century German literature, soon forgetting all about global industry and the art of motorcycles. As a result, prior to coming to Georgia Tech, I spent nineteen years at the University of Kentucky, and just under half of that time as Chair of the German Department. During that time, I "joined" the 20th century, but I did so more or less as a result of my interaction, integration, and fascination with an entirely new German-speaking culture and society in the German Democratic Republic (GDR), where I spent seven summers before the fall of the wall, eventually inviting numerous dissident writers and intellectuals to the US from 1986 to 1992. Perhaps, the fact that literature was so important and deeply integrated into daily life in that country—and functioned as a means of information not available in the state-controlled press—played a role in how I began to see the connections between life and literature in ways that hearkened back to my days at CU.

GDR literature played a courageous role, not generally recognized in the West, with its influence on the social and political mentality of the population who eventually changed the course of history with their protests—and with a little help from Gorbachev. Science, politics, and literature sometimes converged in the GDR, such as in the case of the highly respected chemist Robert Havemann. He was sentenced to death by the Nazis in 1943 for participating in the resistance, and escaping this fate, was later stripped of his professorship at the Humboldt University and his party membership in the GDR in 1964 for his lecture series on "Dialect without Dogma." He soon joined

[11] Storage Tek formed in 1969 by four ex IBM engineers from the plant near Niwot, CO and developed into the top international company in Colorado with booming sales and constant construction of new facilities, then experienced several R&D failures and declared bankruptcy in 1984. They did emerge again successfully, finally selling off to Sun Microsystems for $4 Billion in 2005.

forces with younger writers and musicians, some of whom were expelled from the country, to work for a better society in the GDR.

My friendship with one of the most famous writers of the GDR, Christoph Hein, showed me how literature in the form of intellectual creativity could form a matrix of aesthetics, historical analysis, literature, and political events to demonstrate how people are affected by different aspects and implementations of public policy. Perhaps it seems odd to mention Hein in the context of encounters on the pathway to global engineering, but I have not lost sight of my passion for literature that really is connected to life, and so therefore the question becomes, how can intelligence function to influence policy? Do we want to produce engineers—or other graduates for that matter—who will work quietly on their projects and in their labs producing commercially viable products without regard to their impact on the environment or economic structures, or do we want to produce an intelligentsia that will understand how to employ technology, science, and the humanities in ways that have a principled, positive impact on society and the world in which we live?

TRANSITION TO BUSINESS GERMAN

Equally as important to my development were the abrupt disappearance of my "adopted" country in 1990, and the resulting rapid collapse of the business of scholarly research into the literature, culture, and society connected with its existence. To me, the vanishing GDR was somewhat akin to what may someday occur for engineers in the automotive industry with, say, a sudden but necessary end of combustible fossil fuel engines. They will adapt and adjust, whether to hydrogen fuel cells, electric cars, or other technology. Unfortunately, the field of literary scholarship retreated into the abstract, into *the l'art pour l'art* cycle that comes and goes from time to time, and its relevance in society has perhaps diminished somewhat due to this trend as much as to how the technology of the internet is pushing good books aside.

Toward the end of my tenure in Kentucky, I began to ask myself whether or not I could in good conscience advise students to major in German. What kind of a job could they expect to get? In the late 1990s, while thinking about strategies to address then-declining enrollments in German in the context of language learning that encompassed not only culture and literature, but science, business, technology, politics, and international relations as well, I was coincidentally contacted by Osram Sylvania. Osram, a subsidiary of Siemens, had just acquired Sylvania Corporation, which had two plants operating in central Kentucky. The German company had shipped some major machinery for the manufacture of halogen lamps, along with a number of engineers to help with installation and operational start-up. The holdover employees from Sylvania were having communication problems with their new counterparts in Germany, with the technical language of electrical engineering, with the German language operations manual for the machinery, and even with the German engineers themselves, whose English was not as good as presumed. There were also problems with the speed and efficiency of the new machinery.

The request from Osram Sylvania was to teach German to management, engineers and staff and to help with technical translation. As I thought about this, I was again reminded of how I had met the Director of Manufacturing for Brunswick Corp., the largest firm in Stillwater, Oklahoma at the time, on the plane to Tulsa for that long-since-forgotten Applied Languages Conference in 1983. He gave me a ride from the airport into Stillwater and told me how Brunswick had contracted to obtain German-made components for an item they were assembling and marketing. The instructions were in German and the translations were inadequate, causing them to pay extra for a group of German engineers to come in to explain the operation and functions to their own engineers. When trade opened with China, he told me, they were one of the first companies to go in with a joint-venture proposal. When closing the deal, they discovered that their Chinese interpreter was unprepared to handle the complexities of the financial and legal terminology in the contract, including key clauses on technology transfer.

I had done a book-length technical translation years earlier of a Swiss energy policy study to determine whether Switzerland should embrace or avoid nuclear power. I did not like the tediousness of the work, which was highly technical in areas where I had little competency. But I agreed to a teaching/consulting contract with Osram Sylvania structured over two years at two days per week late in the working day, with extra fees for technical translation. Although this turned out to be a near-impossible and at times frustrating task because the managers and engineers were called away from the office intermittently for up to two or three weeks at a time, it led to considerable reflection, especially because of the interaction with the engineers, with whom I got along well because of their sense of humor and hands-on attitude.

The experience seemed to encourage other key changes in perspective for me. Meanwhile, during this time I published books and articles on German literature of the 18th century and on Christoph Hein and was promoted to full professor, leaving me with a feeling that I was developing an irreconcilable split in my philosophical outlook on the world, and was perhaps diverting and diluting my energy. On the other hand, promotion has its privileges in terms of freedom to think differently, and so does chairing the department.

I completed some self-training in Business German and began to teach a section each year. At the same time, I became involved in an interdisciplinary program between Economics and foreign languages. I co-taught a class on German economics with a professor of economics, he a native German speaker from Austria conducting his half in English, of course, and me locating authentic sources on the economics topics in German for students and teaching in German. However, what I recall the most from this time is how Economics had lost internship lines set up by my Austrian colleague with Siemens and Mercedes by sending students to them who just could not speak any German. The companies then declined to accept any further students from UK. It was clear to me that we could not afford to send students abroad to work who might perform poorly either from the standpoint of intercultural communication or on the basis of inadequate technical skills.

What I recall most from the Business German class, which usually did not exactly grip my imagination, was the semester I dropped the syllabus in mid-term and adapted the course to a case study of Vodafone's takeover of Mannesmann. Although students had to research the pros and cons of the takeover from an economic standpoint, I was most struck by the dismay in the German media at the prospect of losing an old and highly revered company to the British. Mannesmann was a highly diversified company, but Vodafone was clearly only interested in the telecommunication know-how for its development of D2 technology, as well as the telecommunication partnerships Mannesmann had developed in Italy, France and Austria.

A major negative cultural impact was at the center of the argument for the Germans, who urged the shareholders to not sell out in order to preserve the identity of the company, regardless of the financial benefits. It was a classical German engineering company, formed by Max and Reinhold Mannesmann to exploit their 1885 invention of the *Schrägwalzen* (skew rolling) procedure to build seamless tubing. When Vodafone inevitably acquired Mannesmann, it only kept the telecommunications technology and subsidiaries, selling off Dematic (logistics), Fichtel & Sachs (automotive) and Krauss-Maffei (plastics), leaving a restructured Mannesmann eventually with just its pipe and tubing industry.

The ripple effect was felt in Kentucky where the various subsidiaries were subject to personnel and management changes as Siemens, ZF Friedrichshafen, and others bought up those businesses and reorganized. Although the intricacies of this project had begun to capture my interest, the key moment in class occurred at the end when I asked the students, who by then had achieved a reasonable level of German within the scope of the class, if during a job interview with a company doing business with Germany, they would be able to state confidently that they could represent a US company by interacting in German with counterparts from German companies, be it in marketing, finance, communication, etc. Each student responded, with "no."

PROGRAM DEVELOPMENT

My path continued to crystallize and I developed a passion for creating programs designed to produce language proficiency levels that would enable practical applications of expertise gained. Building such programs could both attract students with excellent skills in their discipline and facilitate partnerships with companies on the basis of students that we could "sell" to them. An internship would be a convenient and cheap way for companies to screen future employees, and it would send students abroad for long enough to build the kind of language and cultural proficiency I believed would be necessary for success. As an incurable dreamer, I also believed the acculturation taking place while abroad would produce in such individuals high ethical standards and a great appreciation for developing mutually beneficial business deals that would not leave the global impression that the US was ruthlessly exploiting other countries for profit, thereby overcoming the "ugly American" syndrome.

To my surprise, I was able to identify approximately fifty German companies operating in Kentucky. I met with an associate of the UK development office, and we developed a long-term strategy to create relationships that could lead to funding for UK and, perhaps, to study abroad scholarship endowments. In the summer of 1998 we literally barnstormed Kentucky. I noticed right away that it was easy to talk to engineers and executives. I had been worried about this due to some sort of preconceived negative notions about elitism and wealth that humanities types develop over the years, and some of them told me they had felt a little uneasy and even intimidated about talking with a German professor as well. The ease of communication was a key factor, and I felt encouraged and energized. I was astonished to hear about the importance companies like Balluff, a mid-size company from Germany that produces electronic sensors, placed in the satisfaction of its employees, with the result that people simply did not want to leave the company because of the positive workplace atmosphere.

One company that did not join our efforts was IT Spring & Wire, which was co-owned by companies from Japan and Germany, both of whom had managers on site who disliked each other and had opposing views on the engineering approach and management of operations. Although I only spoke with the German manager, it was evident from his comments that the cultural chasm between him and the Japanese manager was too wide to overcome, which I did not really understand until my later collaboration with Siemens, during which time I heard more about German difficulties in gaining a foothold in Japan. Germans like to be direct, come to the point and say what they mean. All this produces an image of bluntness and impoliteness with the Japanese, who like to keep their cards more concealed and take a long time to carefully reach a decision. Both cultures are meticulous each in their own way, and both do place substantial weight on preparation, theoretical foundation, and environmental and social impact. But their methods of communication are diametrically opposed.

We visited all the Mannesmann companies in Kentucky I had read about, and I continued discussions with Osram Sylvania. Many of the companies wanted to know more about their counterparts. These conversations inspired a meeting with the Kentucky Economic Development Office, and we began to discuss organizing a group. We hosted a conference on international business at UK the spring of 2000, with papers and presentations from the Governor's Office, several of the company representatives, and UK faculty. The publicity attracted the attention of the German-American Chamber of Commerce, whose CEO attended from the home office in Chicago. He offered to serve as an umbrella for our organization at discount rates for its members that would open access for them to national and international networking opportunities and provide additional publicity. Many of the company executives believed that the organization would provide them leverage at the State government level. Twenty-three of the fifty German-owned companies agreed to join what we named the Kentucky-Germany Business Council or the "KGB" Council as we liked to call it. All but two of the companies we had visited agreed to join and the CEO of Osram Sylvania was named the first president. By 2000, internship opportunities were opening for students in both engineering and the Foreign Language and International Economics degree program, and I was looking into a larger future.

SCALING UP AT GEORGIA TECH

My intellectual and practical development had prepared me well for the move to Georgia Tech in the fall of 2001, probably the easiest decision I ever made. According to my preliminary research, the scale to develop similar opportunities was exponentially greater and could be expanded to opportunities around the globe into many countries involving numerous languages. The Atlanta area had some

1600 foreign owned companies, of which over 300 were Japanese, another 300 were German, and around 130 French-owned.

In addition, the job consisted of chairing the School of Modern Languages with its eight languages and cultures. During my campus visits, I realized the uniqueness of Georgia Tech as a place where the traditional barriers between academic units were essentially absent, where collaboration and entrepreneurism was highly encouraged. I had often experienced, when I thought I had a good idea, reactions by colleagues and administrators who usually explained why it was not feasible. But at Georgia Tech, good ideas are met with enthusiasm, although it is generally necessary to find your own funding to implement them. Perhaps this is attributable to a couple of civil engineers who ran the place: Wayne Clough, whose vision of collaborative entrepreneurial engineering included liberal arts and internationalizing education, and Jean-Lou Chameau, who often used to speak about the concept of the "renaissance engineer."

Georgia Tech was at the forefront of internationalization with a goal for 50% of undergraduates to complete an experience abroad. I was impressed by a statement in the strategic plan that reflects on the collaborative potential between liberal arts and engineering and differentiates us from many other technological universities: "Just as engineering has recognized the ways that its work is permeated by disciplines represented in the liberal arts, the liberal arts have become transformed by their own engagement with technology," creating a national and international "model for educational change."[12]

Clough formulated ideas that seemed to merge with my philosophical questioning, posing at one time the question: "Why should those who may not understand technology be the arbiters in deciding how it will be used, or whether it will be used to a good or evil purpose? Are we only about the objects and practical ends … of technology or should we extend our abilities to broader ends?" Clough was leading into his idea of "vanishing boundaries" that projected an education that is collaborative, interdisciplinary, and global, one that would lead to an opportunity to become "a citizen of the world as well as an architect, scientist, engineer, historian, or business leader." Clough stated that the lack of communication between technology and the humanities is deeply rooted, creating "a major hindrance to solving the world's most significant problems."[13] Since Clough and Chameau were engineers the statements were not idle platitudes, on the contrary, the mandate was to implement without "piddling around," as Jack Lohmann (Vice Provost for Institutional Development) later put it in reference to the International Plan.

My first task was to develop the newly implemented joint major in Modern Languages and International Affairs, which grew spectacularly from four majors to 185 in about four years. I also collaborated immediately with Economics to develop the new Global Economics and Modern Languages joint degree. But the real journey at Georgia Tech began with two chance encounters: first a meeting with Christian Callegari, at the time in charge of North American recruiting and university relations for Siemens Corporation, and the second with officials from the Japan Export

---

[12] Georgia Institute of Technology, "Defining the Technological Research University of the 21st Century," 2002, 7.
[13] Clough, Vanishing Boundaries, 2007.

Trade Organization (JETRO). Callegari was a charismatic, dedicated, and aggressive organizer who sought to create a consortium of international universities to converge with the research and development interests of Siemens. Someone in engineering had suggested he meet with me when he was visiting in the fall of 2002. Imagine an engineer at any other university suggesting that someone like Callegari should drop in on the language department? We spoke for an hour or two in German. The encounter led quickly to a trip to Munich for visits with the Technical University of Munich (TUM) and with Siemens corporate headquarters.

From this, emerged the Georgia Tech-TU Munich-Siemens model for internationalizing the curriculum. This model became for me a prototype for successfully combining engineering, etc., with disciplinary practices abroad to accomplish a professional level of foreign language and intercultural competency. The model includes two years of in-house language preparation, our signature faculty-led summer intensive Language for Business and Technology (LBAT) program, a study semester at TU Munich taking engineering courses taught in German, plus a five-month internship in Germany. For JETRO, I became the regional program manager to place GT students for six to twelve month internships with Japanese companies. With the help of the Japanese unit, we would send almost as many students to Japan as to Germany in the next five years.

I should not neglect to mention that during this time I taught a course on German drama and lyric at Georgia Tech, which attracted mostly engineers, perhaps seeking some relief from the uncompromising rigor of engineering classes. The Goethe Institute in Atlanta invited regional universities to bring their students to an event in which they would present readings and skits from German poetry. Once again, I dropped the class syllabus, and we decided to perform what we called "Faust for the 21st Century." We read Faust I, in which Faust sells his soul to the devil, who makes him young again so that he can fall in love with an underage girl, who gets pregnant. Since German laws in the 18th century placed the entire blame on the woman in such cases—and Mephistopheles astutely diverted Faust's attention in the critical hour when he could have rescued her—she was subject to imprisonment for getting pregnant without being married. In her despair, she turns temporarily insane and kills the baby, for which she is summarily executed. The damnation for his part in Gretchen's tragedy is what would have been Faust's fate had he not been redeemed by becoming an engineer many years later (Part II).

A natural talent, MacField Young (ME) played Faust. The students were responsible for re-structuring the plot into fifteen minutes, selecting and condensing the parts that would re-interpret the play for the 21st century, and learning how to convey their interpretation. The engineering students were much more creative than the liberal arts majors, although the student playing Gretchen came up with some excellent music. The engineers decided to split Mephistopheles into two persons, each speaking into one of Faust's ears, which had a brilliant effect. I decided that we would not just read it, but that they needed to learn the script and rehearse it so that the concept they had created would be communicated effectively. Plus, we wanted our show to be the best. The event left a lasting impression on me and formed my idea of what an excellent engineering student could do and be in the larger picture of life. Although I am not qualified to comment on their engineering skills, their successes will likely be measured as much on the strength of their ability to articulate, be creative, and operate in many unusual environments as on their engineering skills.

The TU Munich/Siemens project gained some recognition at Georgia Tech, eventually leading to research and graduate degree collaboration with TUM. Callegari insisted that I attend the International Engineering Colloquium at the University of Rhode Island. I had heard of John Grandin, of course, a fellow German professor, and I discovered that his path had preceded mine by many years, setting a precedent of substantial importance for those of us who followed. I was most struck during that first of many conferences I attended, however, by the keynote speaker from BMW, Brenda Cox, Manager for International and Advanced Procurement from the plant in South Carolina, who said that when management came to Spartanburg the meetings were held in English, but when US management was called to Munich, the meetings were held in German. As a result, she had taken it upon herself to learn German, and to learn it well.

Callegari later suggested that the International Engineering Colloquium take its message on the road, hoping to have it alternate between Rhode Island and other universities with which he was building collaborations, John agreed and Georgia Tech hosted the Colloquium in November 2005, just as the International Plan was finalized and implemented. I am sure I could never have predicted that I would be co-organizing a conference on international engineering, but with key help from Jack Lohmann, who helped organize some of the panels, and from John and his staff, we were highly successful.

I was able to invite Paul Camuti, President and CEO of Corporate Research for Siemens in North America and Alex Gregory, President and CEO of the first Japanese company to locate in Georgia, YKK. Both gave speeches that I still quote when recruiting students into the International Plan. Speaking to what engineers need in the 21st century Camuti mentioned requirements that have perhaps become standard in the industry, including "business acumen and sense of entrepreneurship, capacity to handle complex systems, cross-cultural sensitivity and social awareness," and added to this "good communication skills, including multiple languages."[14] Alex Gregory spoke of communication and value, stating that "beyond our needing foreign language as a communications tool in conducting international business, foreign language studies gives us insight into the way different people in different cultures actually think and act. *It is the key to understanding not just what is said with words, but what is truly meant*."[15]

THE INTERNATIONAL PLAN COALESCENCE: CHANGING THE IDENTITY OF THE "REAL ENGINEER"

The TUM-Siemens model became one of the basic formulas for the design of the International Plan. The collaborative effort at Georgia Tech began to take shape at the intersection of an accreditation review of the Institute and its intrinsic Quality Enhancement Plan (QEP) and the directive by Chameau to implement a strategy to develop global competency, not only for engineers but for all

---

[14]Camuti, "Engineering the Future," 2005.
[15]Gregory, "The Case for Global Engineering Education," 2005.

disciplines. Jack was charged with preparing the accreditation report and the QEP's aim to strengthen global competency for our students. The project was being formulated at the upper administration level, and the shape and nature of the International Plan immediately went from "piddling around" to "let's just get it done." I served on the two steering committees, one addressing the content of the IP and one addressing work abroad.

The IP conceptualization, which defines "global competency," proceeded without as much difficulty as one might have imagined. The student-learning outcomes included requiring graduates of the International Plan to "demonstrate (1) proficiency in a second language; (2) knowledge about comparative international relations, the world economy, and the socio-political systems and culture of at least one other country or world region; (3) an ability to assimilate comfortably in at least one other country or world region; and (4) knowledge of the practice of their discipline in an international context."[16]

The IP regards education to be competency-based; and global competency "is the product of both education and experience."[17] The combination of language acquisition and foreign experience is therefore integrated into the discipline to produce graduates who can readily use language skills to "comfortably assimilate within other cultures" and "to collaborate professionally with persons in multicultural work environments."[18] This pragmatic-engineering approach may not have addressed the philosophical issues I had in mind, nor those articulated by Clough, but the immersion experience abroad would impact the students' mentality.

The committee was more conservative than I thought it should be, setting a graduation rate of 300 IP students to come on line by the end of the five-year plan. My most significant contribution may have been surveys I had been conducting of students in foreign language classes. I felt direct contact with this group would provide the best feedback and recruiting potential and knew that 65% of the students hailed from engineering and computing. The 2005 survey, shown in the figure below, was incorporated into the document published by Jack Lohmann, and demonstrates—even with a fairly low return rate of less than 50%, that over 400 students were interested in the IP before it even began, and pertinently describes their stated preferences in how to structure the experience abroad, including which countries they would most likely want to visit for an immersion experience.

The fall 2008 and ensuing surveys were calibrated to eliminate "casual" responses with more focus on underclassmen, and it showed steady interest at somewhat higher levels than the 2005 data shown in Figure 14.1. The language numbers have increased dramatically, and the top majors by discipline with IP participation include the joint degree International Affairs and Modern Languages, Electrical & Computer Engineering, Industrial and Systems Engineering, Mechanical Engineering, Management, Computer Science, Aerospace Engineering, Architecture, Biomedical Engineering, Chemistry, and Civil Engineering. The 2008 top destinations desired by students planning to enroll in the IP were France (118), Spain (sixty-four), China (sixty-three), Mexico (forty-seven), Japan (forty-four), and Germany (thirty-one). This ratio has remained the same throughout with only

[16]Lohmann, "Strengthening the Global Competence and Research Experience of Undergraduate Students," 2005, 12.
[17]Lohmann, "Strengthening the Global Competence and Research Experience of Undergraduate Students," 2005, 12.
[18]Lohmann, "Strengthening the Global Competence and Research Experience of Undergraduate Students," 2005, 19.

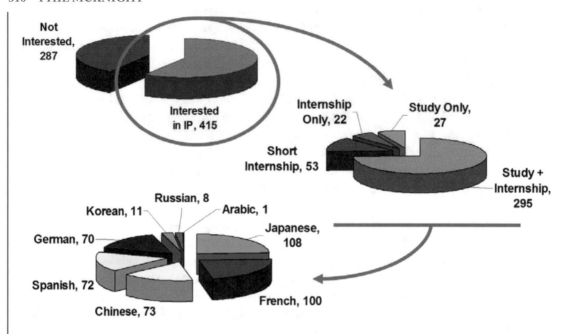

Figure 14.1: Student Interest in the International Plan.

slight variations in the distribution percentages, although the survey has become more efficient and now also typically includes over 2000 responses. Most importantly, 756 students had enrolled in the International Plan as of spring 2010, still quite a bit below the level needed to graduate 300 per year, but recruiting the past two years has increased dramatically, and so has student participation in study or work abroad, regardless of their participation in the International Plan, indicating a general cultural shift in student attitudes.

A still unresolved issue is how to design programs that will not just attract students, but that focus specifically on meeting their needs and preferences, as indicated in the data. We know that Japan is a strong draw for Computing students, that many Mechanical Engineering students have strong interest in Germany, etc., but few disciplines have designed programs specifically in the three or four countries desired by their students. Plus there is not enough coordination between research-oriented faculty, unit advisors, and the Office of International Education to approve programs in all the countries targeted by students.

Although I argued against the inclusion of a two-year language requirement for English track IP students on the basis that we did not want reluctant learners in our classes and that their experience in an English-speaking country was unrelated, the committee overruled me. Approximately 80-90% of the students are now in the foreign language track. They need to achieve "Intermediate High" on the standard ACTFL scale, tested by an outside testing agency. I argued in favor of "Advanced Low," which more directly corresponds to the skills needed to produce students with global competency. The consensus, however, was to be as inclusive as possible and not discourage students at the outset because of the foreign language hurdle.

Although the current (Spring, 2010) data represents only the small number of students who have already completed the IP, of the eighty-nine who have taken the required oral proficiency interview (OPI) thus far, sixty-two have passed and twenty-seven have failed.[19] Thirty-seven achieved Advanced Low or higher, a strong indication of the value gained for students who spend adequate time immersed in either work or study abroad. Nevertheless, the issue of how much foreign language is necessary with the emergence of English as the *lingua franca* in science and engineering has triggered some debate from time to time. Interestingly enough, the principle proponents for reducing the required amount of immersion in a foreign language have come from disciplines such as International Affairs and History, and much less so from engineering and computing. Although students in difficult language countries like China have taken courses in English, the students at TU Munich have all taken their engineering courses in German. English may be used for students studying in Great Britain, Australia, etc., or in conjunction with programs taking place in multiple countries and defined by "intellectual coherence." Nevertheless, the overwhelming majority of students elect a foreign language track.

The IP calls for a minimum of twenty-six weeks abroad in any combination of study, work, and research, for which the GT-TUM-Siemens model is just one path, and most likely one that will not be taken by the majority of students. The evidence shows that most students will achieve the minimum proficiency standard we have established, and the flexibility in creating different models for work/study abroad could permit replication of one or more of these sub-models at other universities that may not wish to duplicate the entire plan.

The major barriers to study abroad are fairly obvious; for students, cannot earn credit in their major, cannot afford it, cannot graduate in four years, other campus commitments, and parents opposed; for the university administration, state legislatures pressure us for four years to graduation, language departments may not cooperate, engineering units may not cooperate, engineering curriculum is too rigid to add more elements, failure to recognize value is added, and foreign academic programs are believed to be below standards.

The major internal obstacle at Georgia Tech consisted of getting the individual disciplines on board. Jack had decided to fly beneath the radar of the Deans as much as possible until the eight-semester plans were ready. For this chapter, I met again with Jack and Howard Rollins, then Director of the Office of International Education, to get their views on what had taken place at this crucial juncture in the planning. Jack felt that the Institute Undergraduate Curriculum Committee needed to be kept out of the discussion in the beginning and that the buy-in from School Chairs was critical. Although Howard initially wanted to create a certificate in global engineering, Jack decided that the International Plan needed to be fully integrated into each discipline, a major and

---

[19] An unusually high number (fourteen) failed in French, due primarily to lack of complete immersion while participating in the School of Architecture's senior design year in Paris, taught in English. The group tended to stay together.

highly significant step in the quality and depth of its design. According to Jack, there were plenty of naysayers at the School level who would contend that "only a real engineer would want to toe the line" and not get involved in study abroad. Why earn credit elsewhere, after all, "aren't we already the best?" Nevertheless, some twenty-six (of thirty-six) disciplines eventually joined.

The curriculum design was a major headache for most Schools, especially to meet the mandate for a four-year curriculum, of which now at least twenty-six weeks would be spent abroad. The models created exemplified "one way of completing the program," which was necessary in order for approval from the Institute's Curriculum Committee. Engineering was not altogether enthusiastic about accepting credit earned abroad. In fact, many were downright skeptical about the value of basic courses from China or even Germany. This problem has not been resolved completely, as some students discover they may not get certain credits when they return, but engineering did indeed provide more flexibility in highly rigid curriculum structures to enable the IP. Trips by faculty advisors to partner institutions abroad helped allay concerns about the quality of engineering in foreign countries, but the objective for students to transfer more credit to fulfill major requirements—not just electives—will require additional time to achieve.

The inclusion of a culminating experience in the form of a capstone course turned out better in practice than in theory when Dave Sanborn (ME) created a senior design course that included students returning from abroad. The students who had interned and studied in Germany contacted Robert Bosch Co. themselves to get the project set up, and were able to successfully integrate their experience abroad into the capstone, and to communicate the differences between the two countries to American students in the course who had not been abroad. Creating more opportunities for work abroad was and is, in my opinion, the key to success for the program since these internships will attract engineering students. To facilitate this, a Work Abroad Office was created to establish relationships with companies and organizations willing to participate.

Howard told me that the President wanted a QEP that could be implemented and achieve its goals, and Jack pointed to Clough's authorization of the three million dollars for the IP as the key for getting engineering on board. Howard realized that "successes depend on a plan that would enable a continuation of support beyond the five years; that would beef up financial aid for students in the program; and prevent backsliding of academic units so they would not fail to maintain a critical mass of students." The three required courses designed to deepen the experience were also "enrollment opportunities for individual units," according to Howard, "that would encourage their buy-in." Funding to each unit to develop capstone courses was made available, and students may apply for a $750 grant when they go abroad. Modern Languages received a substantial sum to deliver courses to the anticipated increased numbers of students.

Jack made a key decision to prevent diluting the QEP by selecting only two of seven proposals submitted by various units. This action created a viable fund that would eliminate "just piddling around." Jack was convinced from the outset that students needed a more global education and felt that although Georgia Tech was already out in front with its campus in Metz, France (Georgia Tech Lorraine, GTL) and other international presences, he wanted to be more "cutthroat" in positioning

Georgia Tech to compete with peer institutions for the best students. Jack envisioned a "cohesive, comprehensive program difficult to replicate," but it would also evolve into just "partly what you do to get prepared" for life while at Tech. According to Jack, the defining moment came on a trip to GTL in a meeting with then GTL president Teddy Püttgen, the executive director at GTL Bill Sayle (deceased 2008), and Bill Wepfer (since 2008 Chair of ME). Bill Sayle quietly took notes during the conversation and scribbled out an eight-semester program for Electrical Engineering, declaring it could be done. The next step was to find "a Bill Sayle in each college and in each program in each college." After that, selling it up the line was easy. None of us, however, really addressed the transformational change that students might experience; we were focused on the logistics of getting a workable structure in place.

REFLECTIONS, PERSPECTIVES, AND THE FUTURE TRANSFORMATIONAL EXPERIENCES

Initiatives from the engineering sector seem to gain much more ground than other disciplines in internationalizing the educational process, notwithstanding the remaining "naysayers" that may be on campus. In Modern Languages, we took pains to identify ourselves as partner, not a "service unit." We structure our programs to prepare students for life and work in the global workforce by teaching advanced communication skills and creative thinking, and by opening access to other cultures in order to develop professional competency in the language. I tend to be uncompromising on this last point, and I continually face efforts to reduce the need for linguistic competency from professional quality to the ability needed to just get along in another language. Sometimes I have compromised on this issue since some ability is usually better than none, and we need programs with variable requirements to meet our students' needs, but I devote most of my resources and energy to constructing pathways with the transformational potential to enable students to become part of another culture and mentality.

Of principal interest from my non-engineering perspective is the educational impact of foreign experiences on any discipline, what difference it might make in the life of students who are able to develop advanced competency in communication, deep intercultural skills, and comprehension of the society in which they conduct their immersion experience. Engineering seeks innovation and new technology, not only in terms of "what holds the world in essence together," but how to build better dams and commercialize them. Global competency could be measured in engineering improvements achieved by graduates integrating methods learned during their international experience. Does our program match the development of students' global competency to the needs of industry described by the executives? My role involves a related part of the process, the part dealing with communication and not technical discovery.

The Office of Assessment at Georgia Tech now surveys all incoming freshmen with intercultural competency indexes. A comparison between those who participated in an extensive experience abroad with those who did not may provide some answers, and Tech would like to determine what the most effective practices for achieving global competence are. Engineers evidently had previously believed that "preparation for the global practice" was not necessary. But recent studies conclude that the "international preparation of engineering students is not just a matter of cultural awareness" gained in add-on pieces of the curriculum, but "it is a matter of professional competence in a global context," and as such providing opportunities "to live, work, or study in a foreign culture is critical to the development of global engineers."[20] Studies like the Lincoln Commission Report have already made conclusions that will emerge from assessment, although details and specifics may vary. We know that students who have traveled, worked, and studied abroad are more "confident and committed to their educational purposes" and have more "poise, self-esteem, autonomy, self-confidence, flexibility, maturity, self-reliance and improved social skills," making them more competitive "in today's 'flat' world."[21]

We think that most of the engineering students returning from abroad will meet goals desired by industry and their own aspirations to be competitive in the job market. But how does experience derived from another language and culture impact or change identity and values? If there are greater things to be gained than enhanced career opportunities and potential increased earnings, what are they? Are leadership skills emerging as well? What about a philosophy of life? How far along the developmental stages to intercultural sensitivity have they progressed? Have they become not only an engineer but also a citizen of the world?

As mentioned, Georgia Tech has begun to survey incoming freshman[22] with such tools as the National Survey of Student Engagement (NSSE), the Cooperative Institutional Research (CIRP) survey, and the Intercultural Development Inventory (IDI), based on Milton J. Bennet's Developmental Model of Intercultural Sensitivity[23] (DMIS). The self-selecting students in the International Plan demonstrate a significantly higher ability to understand people of other racial or ethnic backgrounds even before going abroad. The CIRP results show higher scores for IP students across the board in categories they consider as essential and important, such as: Influencing the Political Structure, Participating in a Community Action Program, Improving my Understanding of other Cultures, Influencing Social Values, Being Involved in Cleaning the Environment, Becoming a Community Leader, and Developing a Meaningful Philosophy of Life.

The IDI is more complex and truly requires time spent immersed in a foreign culture and language in order to pass through the "Ethnocentric Stages" (denial of cultural differences, defense of us vs. them, minimization of cultural differences) to "Ethnorelative Stages" (acceptance of cultural differences in context, adaptation and empathy to other cultures, integration and true bicultural or multicultural attitudes). Research places females ahead of males in all stages of the IDI. At Georgia Tech (191 voluntary samples, with just nineteen IP students returning from abroad for this preliminary study) 48.4% of males in the IP had reached the level of "minimization," vs. 38.7% not in the IP. For IP females 54.3% had reached "minimization," vs. 51.2% not in the IP. Outcomes for

---

[20] Anderl et al., *In Search of Global Engineering Excellence*, 2006, 41.

[21] Lederman, "Quantity or Quality in Study Abroad?" 2007.

[22] For the initial reports from these surveys see Lohmann et al., "Annual Impact Report of the Quality Enhancement Plan on Student Learning," 2008.

[23] See Bennet, "Towards a Developmental Model of Intercultural Sensitivity," 1993.

students in the IP returning from abroad should show significant results once there is more data. The IDI does, in my opinion, measure how well students may be able to interact constructively in other societies, and perhaps the CIRP also is a reasonable indicator of whether a student could be a global and responsible engineer. But we need to know how IP students develop during the first ten years after graduation and whether over the long haul their careers reflect the values inherent in the study/work abroad experience.

My own "data" on student experiences is mostly anecdotal. In Modern Languages, we pre-screen students for the two programs in Germany and Japan. We try to exclude students whom we subjectively feel are not adaptable to foreign cultures with interviews by faculty in the target language. The extremely low failure rate for these two programs substantiates this practice, but Georgia Tech students bring with them a reputation for quickly adapting to workplaces and contributing to the projects they work on. It is rarely different when they go abroad. All but one or two reported that their language skills had improved tremendously and that this made a major difference in their ability to socialize and to professionally interact with their team members to achieve project goals and integrate culturally. They created lifetime friendships, and they adapted to major differences in workplace procedures. Most returned determined to find a position that would enable him or her to apply their linguistic and cultural skills on a life-long basis.

Not every student succeeds, nor would it attest to the rigor of the program if they did. Many students drop out because of the extra work or because of the difficulty of their engineering program. The steering committee had estimated that attrition could be as high as 50%, and this is proving to be accurate. The oral proficiency test in the chosen language has been challenging for those who should have either immersed themselves more thoroughly, or should have stayed longer than the minimum required—particularly true of languages like Chinese or Korean. Some drop out due to the extra course load required and some because they are satisfied with their international resume and do not feel a need to complete the last requirements.

There are many success stories, however. The first three students who participated in the TU Munich-Siemens program (Nick Karnezos (ME), Doug Niggley (AE), who actually did his internship with Lufthansa Technik, and my *Faust* actor Mac Young (ME), who after his internship with Siemens did a second internship with Bosch, only this time in Brazil, to improve his Portuguese) made an impressive presentation[24] at the 2005 International Engineering Education Colloquium at Georgia Tech. That spring Jack, Howard, I and others presented the International Plan to the Georgia Tech Advisory Board. After we professors droned on for some time, Doug and Mac took the floor and basically brought the house down, so to speak, with executives in the audience, including Aubert Martin, at the time CEO of the Siemens Power & Automation plant in Alpharetta, spontaneously jumping up to exclaim, "We need more students like them!" Siemens went on to create an "International Associates Program," recruiting outstanding graduates with excellent German skills to be placed for two years in Germany.

---

[24]Cf. Young et al., "Georgia Tech-TU Munich Exchange and Internship Program," 2005.

These students would then return to Siemens facilities in the US in a management track, and with the skill set to bridge communication gaps that result from differences in workplace cultures and problem solving approaches, thereby meeting the criteria set forth by Paul Camuti for global engineers. Doug was hired in the first year of this program, and reported that he was able to conduct his final-round interview onsite in Germany entirely in German, which gave him a distinctive edge over the other candidates. Mac was heavily recruited by Siemens, but chose another company that also offered him a chance to work globally. Nick went to Georgetown University to study international patent law. The success of these and other IP students has been instrumental in reaffirming the value of integrating international work experience, professional competency in a foreign language, and excellent engineering skills into the educational curriculum. And Georgia Tech already reports many students who enroll here because of the IP.

One student I should mention, Eric Johnson, interned twelve months with Yamatake Corporation in Japan. Eric struggled for two months in Japan as his colleagues and supervisors labored over the details of the project, the composition of the team and how he would fit in, and he struggled with business and technical Japanese, and with trying to adapt to the management and workplace style of the company. But he reported that once the project got up and running, his work on software development and integrated research platforms and the acquired ability to accomplish project goals within a company team operating mostly in Japanese led him to feel confident and comfortable enough to even appear on Japanese television to discuss his experiences. In keeping with the Japanese saying that, "One who never climbs Mount Fuji is a fool, and one who climbs twice is twice the fool," Eric climbed Mount Fuji (once) and became thoroughly immersed in Japanese ways of thinking. Because they were impressed with Eric, the company contacted us to arrange a visit to Georgia Tech to explore potential R&D collaboration for their measurement and control systems and industrial automation sectors. The relationship between a technological university and industry had come full circle on the strength of a student's performance that demonstrated not just technical skill, but the ability to adapt thoroughly to another culture, to a considerable extent by going through the window of its language and culture.

I had formed a friendship with Peter Olfs, retired as the Senior Director in Corporate Communications for Siemens, and he was a regular speaker at Georgia Tech and on the International Engineering circuit for many years. Although he had usually stressed the need for learning at least one foreign language, towards the end of his speaking tours, I heard him comment about the universal use of English and the possibility of going global with just a minimal knowledge of a second language. I asked him if he thought that we would have formed such a close relationship with Siemens, and would have developed the Siemens-TU Munich program as successfully had I not spoken fluent German with him, Callegari and other Siemens executives and employees. He responded, "No."

REFERENCES

Anderl, Rainer et al. *In Search of Global Engineering Excellence*. Hanover: Continental AG, 2006. 314

Bennet, Milton J. "Towards a Developmental Model of Intercultural Sensitivity." In *Education for the Intercultural Experience*, edited by R. Michael Paige, 179–196. Yarmouth, ME: Intercultural Press, 1993. See also the outline of the IDI online at `http://www.library.wisc.edu/EDVRC/docs/public/pdfs/SEEDReadings/intCulSens.pdf`. 314

Camuti, Paul. "Engineering the Future: Staying Competitive in the Global Economy." Keynote Speech presented at the 8th Annual Colloquium on International Engineering Education, Atlanta, GA, United States, 2005. 308

Clough, G. Wayne. *Vanishing Boundaries. State of the Institute Speech, Georgia Institute of Technology*, October, 2007. `http://smartech.gatech.edu/handle/1853/22240`. Accessed August 25, 2010. 306

Geisler, Michael et al. "Foreign Languages and Higher Education: New Structures for a Changed World." Ad Hoc Committee on Foreign Languages, Modern Language Association, 2007. `http://www.mla.org/flreport`. Accessed August 25, 2010. DOI: 10.1632/prof.2007.2007.1.234 299

Georgia Institute of Technology. "Defining the Technological Research University of the 21st Century. The Strategic Plan of Georgia Tech." Atlanta: Georgia Institute of Technology, 2002. 306

Gregory, Alex. "The Case for Global Engineering Education." Paper presented at the 8th Annual Colloquium on International Engineering Education, Atlanta, GA, United States, 2005. 308

Gross, Patrick W., William W. Lewis, et al. "Education for Global Leadership: The Importance of International Studies and Foreign Language Education for U.S. Economic and National Security." Washington, DC: Committee for Economic Development, 2006. `http://www.ced.org/library/reports/40/208-education-for-global-leadership`. August 25, 2010. 298

Gutierrez, Robert et al. "The Value of International Education to U.S. Business and Industry Leaders: Key Findings from a Survey of CEOs." Institute of International Education Briefing Paper, 2009. `http://www.iie.org/en/Research-and-Publications/Publications-and-Reports/IIE-Bookstore/Value-of-International-Education`. Accessed August 25, 2010. 298

Hudzik, John C. et al. *The State and Future of Study Abroad in the United States: A Briefing Book for the Bipartisan Commission on the Abraham Lincoln Study Abroad Fellowship Program*, 2004. `www.cob.sjsu.edu/gem/lincolnbriefingbook.pdf`. Accessed August 25, 2010. 295

Institute of International Education (IIE). "U.S. Study Abroad up 8%, Continuing Decade-Long Growth." Open Doors Network, 17 November 2008. `http://opendoors.iienetwork.org/?p=131592`. Accessed August 25, 2010. 295

Institute of International Education (IIE), "Open Doors 2009 Report on International Educational Exchange," section on Fields of Study, 2009. `http://opendoors.iienetwork.org/?p=150836`. Accessed August 25, 2010. 297

Lederman, Doug. "Quantity or Quality in Study Abroad?" 2007. `http://www.insidehighered.com/news/2007/02/21/abroad`. Accessed August 25, 2010. 314

Lohmann, Jack et al. Annual Impact Report of the Quality Enhancement Plan on Student Learning: Strengthening the Global Competence and Research Experiences of Undergraduate Students. Atlanta: Georgia Institute of Technology, 2008. `http://www.accreditation.gatech.edu/wp-content/uploads/2009/09/QEP-SACS-Annual-Impact-Report-2007--2008-FINAL-WEB.pdf`. Accessed August 25, 2010. 314

Lohmann, Jack. "Strengthening the Global Competence and Research Experience of Undergraduate Students."Atlanta: Georgia Tech Institute Communications & Public Affairs, 2005. `www.assessment.gatech.edu/wp-content/uploads/QEP.pdf`. Accessed August 25, 2010. 309

Marginson, Simon. "The Rise of the Global University: Five New Tensions." *The Chronicle of Higher Education,* 30 May 2010. `http://chronicle.com/article/The-Rise-of-the-Global/65694/?sid=pm&utm_source=pm&utm_medium=en`. Accessed August 25, 2010. 296

McKnight, Phil. "Foreign Language Study: Bankruptcy and Merger." In *Applied Language Study: New Objectives, New Methods*, edited by John Joseph, 19–29. Lanham, MD: University Press of America, 1984. 296

McPherson, M. Peter et al. "One Million Americans Studying Abroad: Global Competence &National Needs." Washington, DC: Commission on the Abraham Lincoln Study Abroad Fellowship Program, 2005. `www.aplu.org/NetCommunity/Document.Doc?id=190`. Accessed August 25, 2010. 298

Obst, Daniel et al. "Meeting America's Global Education Challenge. Current Trends in U.S. Study Abroad & The Impact of Strategic Diversity Initiatives." New York: Institute of International Education, 2007. `http://www.iienetwork.org/file_depot/0--10000000/0-10000/1710/folder/62450/IIE+Study+Abroad+White+Paper+I.pdf`. Accessed August 25, 2010. 298

U.S. Department of Education and the Center for International Studies at the University of Chicago Symposium. *Environmental Challenges across Asia.* March 2, 2007. 297

Wagenhofer, Erwin (Director). *We Feed the World.* Ein Film (DVD), 2005. `www.essen-global.de`. Accessed August 25, 2010. 297

Young, MacField, Doug Niggley, and Nicholas Karnezos. "Georgia Tech-TU Munich Exchange and Internship Program." Presented at the 8th Annual Colloquium on International Engineering

Education, Atlanta, GA, United States, 2005. http://www.modlangs.gatech.edu/degree-programs/international-plan/StudyAbroadExperience_hi.wmv. Accessed August 25, 2010. 315

CHAPTER 15

Bridging Two Worlds

John M. Grandin

What is the likelihood that a professor of German, an author of several articles and a book on Franz Kafka, would become an authority on the globalization of engineering education? What is the likelihood that a professor of German would be able to encourage hundreds of engineering students to become proficient in a second or third language and spend a year of work and study abroad? What is the likelihood that a professor of German would develop a network with dozens of global companies to ensure internships for students and long-term positions for graduates as well as significant financial support for the overall programmatic effort?

As I look back over the twenty-three years of my work as developer and director of the International Engineering Program at the University of Rhode Island (URI) and consider the paucity of similar developments elsewhere, I have to conclude that the likelihood of any of these is not too high. Yet, this has been the direction of my career since 1987, and it is the journey I describe in the following pages. My hope is that my description of this unlikely but very rewarding pathway across the disciplines will provide some insight into the nature of innovative and entrepreneurial academic program building for the benefit of other aspiring faculty and their students.

AN UNEXPECTED NEW DIRECTION

People sometimes ask if I am German or if I have German heritage, in search of my reason for becoming a German professor and international educator. But my path to that career was driven neither by ancestry nor even deliberate planning. It harks back, first of all, to advice from my clergyman father shortly before I was to start my freshman year at Kalamazoo College, a small liberal arts college in Michigan. Since he envisioned me following in his footsteps and knew I was interested in philosophy, he advised me that learning to read German would be sound preparation for studying the great philosophers and theologians. So I signed up for German 101, even though Latin in high school had been disastrous for me and the thought of another language was really not at all appealing. To my surprise German turned out to be my favorite course during my freshman year, largely because I really admired the instructor. Thus, I was happy to enroll in intermediate German for my sophomore year, not knowing what a decisive step that would be. Unbeknownst to me, one of the Kalamazoo College trustees had just given the school two million dollars to create an endowment to support experiences abroad for its students. I was thus shocked when my German teacher, Frau Elizabeth Mayer, asked me after class one day if I wanted to spend the coming summer in Germany, living with a family in Bonn and taking an intensive German course at the university.

As the twenty-year-old son of the Baptist minister from a small, homogeneous New England community in central Massachusetts, being sent to Germany in the summer of 1960 ushered me into a very different world. I learned, first, that there were people who actually spoke German, and that I too could learn to communicate in that language and have fun doing it. Next, I learned there were lots of very good people who had been shaped by very different experiences and had insights into history, society, and life quite different from my own. My host "parents" had lived through both world wars and had been driven out of their homeland in Eastern Europe and forced to start a new life in the West. While their first motivation for taking me into their home that summer was no doubt their need for additional income to help support themselves and their four children, their curiosity about young people from America was a strong second reason. There was thus much lively exchange of views and opinions between us during this time. Living in their home for three months, sitting with them at their dinner table every day, watching how they interacted as a family, hearing their concerns and aspirations, finding myself and my worldview challenged by them on a daily basis, the world took on new dimensions for me in that short summer, as it did for them as well. I found many of my own basic values and political assumptions challenged and began to understand, for example, that being an American was taken quite casually by my friends and myself and that most of us were totally unaware of the life experiences and priorities of people in other parts of the world, or of their expectations and hopes of us. Combining this with my travels on a $50 motorcycle through Germany, Belgium, Holland, Switzerland, and Austria, and long discussions in youth hostels with students from numerous countries, as well as my three days in East and West Berlin a year before the Wall was built, I went back home at the end of August as quite a different person.

My summer experience was powerful enough to convince me to take a year off between my Junior and Senior years in college to go back to Germany for a full academic year. This gave me time to learn the language well, to immerse myself deeper in the culture, to travel extensively, and to sort out what I really wanted to study and to do with my life. Even though I would complete my philosophy degree, I had to let my Dad down about his vision of me as a protégé in the ministry. I just was not firm enough in my beliefs to be able to ascend the pulpit on Sunday mornings and offer parishioners what I thought they would want to hear. Aside from that, I did not feel comfortable speaking in public, and I still don't after forty-five years in higher education! Nonetheless, my family background and upbringing have profoundly influenced my own attitude toward my profession and career. The fact that I like to work hard, that I have a strong service orientation, and that I need to feel that I am helping others to shape positive values and meaning for their lives, harks back to my childhood as the son of a Protestant clergyman. Though I am not an active churchman, I nevertheless feel driven by what I do and take it with the utmost seriousness, perhaps even with a missionary zeal, and I owe that passion for service to my Dad.

Though uncertain about career directions, I was, however, more and more convinced that Americans were extremely provincial and culture-bound in their understanding of the world, and that I could perhaps make a difference in this area through some kind of international work, perhaps even teaching. So, even though I graduated with a major in philosophy, I decided to build on my knowledge of German and Germany and thus applied for and won a fellowship for the Master of Arts in Teaching Program at Wesleyan University in Connecticut. Wesleyan was a great intellectual experience for me and also provided me with a full scholarship to spend another year as a student in Germany, to be followed presumably by a career in high school German teaching.

The second year in Germany offered me greater depth of knowledge and better skills in German, as well as new personal contacts and a growing lifelong commitment to promoting and sponsoring dialogue between the United States and Germany. Though I had never really thought of myself as an academic, I then landed a two-year lecturer position at Union College in Schenectady, New York and thus had the luxury of teaching some bright young kids and learning that I could impart to them not only knowledge of the language, but the importance of understanding the perspectives of people from nations other than our own. This was a powerful and formative time for me, which gave me the confidence to teach and believe I could make a difference through my role as a potential professor of German. The two years went by very fast and made it obvious that graduate school was in order since I would need to have a Ph.D. if I really wanted to follow this path.

The next step was a quick three years at the Department of German at the University of Michigan. I knew that the degree would be in German literature, so I had a lot of reading to do and a lot of catch-up work in terms of literary theory and analysis. But that was fine with me, even though my first motivation was to teach the language and to figure out ways to get students over to Germany for experiences such as I myself had known. I enjoyed my research on Franz Kafka and became thoroughly immersed in my dissertation. Yet, in the back of my mind, there was that gnawing feeling that this was a detour around the work that I would consider to be most important.

The summer before entering the Wesleyan masters program had led to much more than a closer connection to Germany. It also meant I would have a mate who would be happy to go off to Germany for sabbaticals and other special occasions abroad over the coming decades. Unaware of the long-term implications, Mrs. Ruth Wilson had contracted me that summer to tutor her daughter, Carol, who was preparing to study in Vienna for her junior year. This marked the beginning of a relationship that led to marriage two years later just before I started my teaching job at Union College, now forty-five years ago. Carol completed her undergraduate degree at Union and then supported us both while I earned my Ph.D. at Michigan. We left Ann Arbor in 1970, with our infant son, Peter, to start a new life in Rhode Island and my forty-year career (or careers) at the University of Rhode Island.

ENCOUNTERING ENGINEERING

Having gone into the field with hopes of attracting lots of students to study German, to study abroad and to incorporate the language learning process and cross-cultural experiences into the core of their very lives, I was soon disappointed to encounter a low level of interest, indeed a growing hostility toward language learning and language study in American academia of the 1970s and early 80s, attributable in part to a new separatism in the wake of anti-Vietnam War backlash. The number of German learners in American colleges and universities was shrinking and fewer and fewer students were interested in German for the sake of studying literature in depth, as I had done. My own situation was made even worse by some of the myopic thinking of colleagues who were entrenched in their own rigid pedagogical modes and did not want to encourage sending students abroad since it would mean even fewer students in their own classrooms. In short, I discovered rapidly that something needed to change in the profession if language learning were to play a meaningful role for American undergrads, and that, for my own sake and personal well-being, I needed to create change or leave the profession for other ventures that might be more fulfilling. There were times in my early career at URI when I was very depressed.

At URI, I was confirming what I had suspected in graduate school, namely, that the traditional literary mission of a language department in American higher education had effectively limited its reach to a very small number of students. Literary studies had become the exclusive focus for advanced courses in U.S. language departments with the adoption of the Humboldt model of university education in the 19th Century, which meant essentially copying or trying to copy the goals and activities of German departments at classic German universities. In my opinion, there was no logical reason for defining literature as the primary and sole end of the language learning process for American students. Being able to study German literature was logically one end product of mastery of the language, but surely not the only one. Though this might sound like a simple and common sense conclusion, the literary tradition was rooted so deeply that its wisdom was not even open for discussion in language circles at that time. To doubt or challenge the literary curriculum was heresy!

Throughout my earlier years at URI, I had quietly initiated a number of outreach programs to other disciplines, particularly, in conjunction with the College of Business. Based on my conclusion that language learning should be related to all fields and not just the study of literature, I encouraged business students and also chemistry students to study German in anticipation of internships with companies in Germany. I had established a relationship, for example, with the leadership of a Rhode Island subsidiary of a German chemical company, began teaching German to their research chemists, and was able to convince them to place a URI chemistry major as an intern at their German headquarters each summer. I found such experimental efforts quite stimulating for both my students and myself and was pleased to be able to arrange such programs abroad for a small number of students. The feedback from both students and companies was excellent, and it became clear that international preparation for professional school students was needed and highly desirable.

Though I had never considered engineering as a potential partner for a department of languages, logic would have suggested such, even at that time. But I did not know the URI engineering faculty, who were located in another corner of campus, and they seldom found reason to interact with humanities faculty. In addition, engineering students had a very lock-step curriculum and were the one group on campus not required to take any language classes at all. Looking back, it is quite clear that there was no obvious or simple way to unite languages and engineering at URI or any other campus for a common cause and that it would take some kind of special circumstance or unusual course of events to get things off the ground.

Coincidentally at that time I, together with my brother, himself an engineer, had built a house next door to my own home on speculation – with the idea of possibly leaving URI and starting a new career as a contractor, such was the extent of my professional discouragement at that time. Given the weak housing market in that year, we had decided to put the house up for rent for a year or so, with plans of selling once the market improved. As fate would have it, we quickly found a tenant in a new hire at URI, who happened to be the new dean of the College of Engineering! His name was Hermann Viets, and he had been born in Germany, raised in a German speaking family, and strongly agreed that engineers would need global preparation if the American technology and economy were to remain strong and competitive in the coming years. He and I thus had a meeting of the minds during a classic American backyard chat, and subsequently formed a committee of language and engineering faculty to lay the groundwork for a new international program for engineering undergraduates at URI.

Though we explored several options, the basic idea of the URI International Engineering Program (IEP) emerged very quickly as the dominant model. It would be a five-year undergraduate program, enabling students to complete both the BS in any one of the engineering disciplines and the BA with a major in a language, initially German. In the fourth year of the program, IEP students would be sent abroad for a six-month professional engineering internship experience. There were doubts whether students would be willing to commit to an extra undergraduate year, and there were doubts whether we could convince faculty on either side of the value of this direction. We knew, after all, that engineering faculty would prefer more time for technical subjects and generally saw little need for language work. And we knew that language faculty tended to be fearful of programs that might possibly suggest their department is in the "service" of a professional school. Languages were, after all, proudly housed in the sphere of the humanities.

Despite the doubters, Hermann Viets and I were very excited about this plan, and we both took it forward with a great deal of personal commitment and passion. The program was able to evolve because of the clear engagement of key persons who were eager to act when given the opportunity, and it certainly made a difference that one of us was a dean. Though I was more than ready and willing to commit my time to this effort, it would have been extremely difficult in the early years without the enthusiastic support of the engineering dean. Hermann's attitude was extremely positive about this venture and he played the key role in selling this idea to the doubters among the engineers, who thought the whole world spoke English and saw no need to "dilute" technology with language and culture study. I, in turn, had my work cut out in convincing colleagues in languages that it was not heresy to develop special language courses for engineers and to move the curriculum in German toward a more comprehensive more inclusive German studies program versus a pure literature major.

It is important to note that the evolution of the IEP coincided very closely with the fall of the Berlin Wall, the opening of China and India to the free market system, the development and spread of the Internet, and the efficiencies of rapid travel. Hermann and I were well aware in the 1980s that things were changing and that business, industry, and technology were becoming global and would demand that future players be able to work internationally and cross-culturally. Our instincts told us that the IEP was important, and this was soon confirmed by the evolving historical context. We had an appropriate idea for the day, at a time when few others in academia were considering these geopolitical changes, the soon-to-be surging wave of globalization, and its implications for the university curriculum. In short, we were ahead of the curve.

GOING OUTSIDE FOR SUPPORT

A critical next step for us involved securing external funding. We knew from the beginning that a new idea such as this would not find immediate support from any regular campus revenue stream. Our colleagues might well tolerate our idea of founding such a program, but not if it took money from existing budgets. In short, we needed money to support new sections of German for the engineers, travel to Germany to line up internship commitments, and develop materials for recruiting. We also needed the prestige of extramural endorsement of our idea in order to help sell it at home. Based on the experiences some colleagues had at other institutions attempting to internationalize professional school curricula, we turned to the Fund for the Improvement of Postsecondary Education (FIPSE), a risk-taking federal agency supportive of new and potentially replicable ideas for higher education at the U.S. Department of Education. FIPSE, with the help of its program officers Sandra Newkirk and Michael Nugent, saw the value of our plans and funded our start-up, actually extending support for our ever-evolving initiatives for a period of more than ten years. Both Sandra Newkirk and Michael Nugent were German speakers and humanists who understood the value of language learning and its relevance to broader educational needs in the evolving era of Globalization. They also understood the myopic thinking of many language faculty and hoped themselves to help launch a new culture for the language curriculum in higher education. The IEP was fortunate to be able to approach FIPSE at a time when its review staff was sympathetic to our cause and open to reshaping the mission of a department of languages.

As a note to those seeking external funding, I would like to stress the importance of persistence and determination. We were not funded the first year, but did go back to the Department of Education a second time, when we were successful. I recall my discouragement at the time due to the highly competitive nature of the FIPSE funding program. But, when asked by a senior colleague in our dean's office if I was planning to resubmit the second year and answering negatively, he literally took me by the arm and sat me down at my Macintosh and insisted that I start writing. Had he not done that, as I myself have done numerous times since then with younger colleagues, the IEP, which has proven itself to be very fundable, may never have become a reality.

Support from FIPSE gave us the prestige we needed on campus and the resources to take the necessary initial steps. We thus launched beginning sections of German for engineers in order to integrate technical vocabulary and culture into the language learning process. We developed promotional materials and started the critical student recruitment program. And we began the corporate outreach needed to ensure that we would have internships for IEP students in their fourth year. Fortunately, our need to do this was confirmed and reinforced by our initial recruitment efforts which yielded over forty enthusiastic students who signed up for beginning German courses to get themselves on track for the five-year BA/BS program, and, especially, the potential six-month internships abroad with global companies.

BUILDING A NETWORK

Despite the clarity of the IEP idea, we also knew from the beginning that the program would require more than the commitment of the faculty from two rather disparate colleges. If we were to place students in professional internships abroad, which was our promise, and if we were to find ongoing financial help for the project, we would have to rapidly develop a partnership with the private sector, convincing them that investment in our students and the program would be to their benefit as well. Hermann Viets brought his considerable skills in collaborating with industry to the table, and I added to that the connections I had made with German-American companies in the area and the German Consulate in Boston. Our common outreach to industry confirmed our belief that the need for globally prepared engineers in the workplace was strong and growing stronger. He and I visited many companies and business leaders together, in both the New England area and in Germany, and this gave me the courage and confidence to take our message to company boardrooms and to ask for support and help by convincing them that our undertaking was of benefit to them as well. This took a bit of time for me as a humanist who was not in tune with the fact that my field really had a market value and that we were on the verge of building something of lasting value, with very broad consequences. Outreach to business leaders, leave alone to other disciplines within the university, was not part of the tool package acquired through graduate school! It was fortunate for the IEP that I was able to take these steps and found the experience exciting and rewarding.

A key action in those early years was the formation of an IEP Advisory Board, comprised of business leaders from the U.S. and abroad, interested citizens, and government leaders, all of whom were equally committed to and passionate about the program. We were very fortunate at the time to be able to convince an influential Rhode Islander to join our effort as Advisory Board Chair. Heidi Kirk Duffy is the widow of one of Rhode Island's most successful industrialists, the late Chester Kirk, who was a URI engineering alumnus. She is a native German, long in the U.S., but nevertheless well-rooted in Germany and well-connected to several key players in German industry. Readily identifying with our cause, Heidi Kirk Duffy was able to help us establish contacts with German companies for internship placements as well as for long-term relationships. At the time of this writing, she has been our only chairperson. She has helped us immensely, both financially and otherwise, and she remains a stalwart supporter and advocate, setting the pace for all of our board members.

Hermann Viets unfortunately did not stay long at URI, as he soon moved to the presidency of the Milwaukee School of Engineering. But he certainly left his imprint on the URI College of Engineering as a whole and did not leave without convincing the key players in his realm of the value of the IEP. Since that time, I have directed the program, including its outreach to the private sector. I make annual visits to German business and industry and have arranged six-month internship placements for well over two-hundred-fifty students. Networking for fundraising purposes has also become a key piece of my annual cycle of activities, and I have successfully brought in several million dollars of support on behalf of the IEP and its students. Today, we speak of our numerous business contacts as partners, many of whom have become avid supporters of the IEP and have reaped the program benefits by hiring our students.

VALUABLE ADMINISTRATIVE EXPERIENCE

My life at URI was complicated at the approximate beginning point of the IEP when the Dean of the College of Arts and Sciences invited me to join his team as Associate Dean for Student and Curricular Affairs. This meant that I would not only be balancing several jobs at once as language teacher, dean, and IEP director, but that I would also be asking myself which direction my career should take. Suddenly, I could consider being a dean, and I did indeed flirt with that idea for a while and even applied for the URI deanship. As I look back, I am grateful that I did not get the position since it no doubt would have ruled out any further development of the IEP.

Though teaching, grant managing, program building, and administering became a tough balancing act, my four years as Associate Dean and one year as Acting Dean nevertheless gave me invaluable experience in terms of making my way through the bureaucracy and politics of the university with the seemingly endless challenges associated with the IEP. In these years, I became acquainted with and generally gained the respect of key people at all levels of the university, thus developing more and more allies for the program. This even enabled me to follow my dean years as the Chair of Languages for six years, working hand-in-hand with administrators who had learned through my incessant proselytizing to recognize the value of applied language learning, i.e., teaching languages for all students and not just humanities majors, and recognizing faculty research in non-literary areas. Becoming Chair of Languages made me realize that the IEP effort and the idea of applied language learning were paying off. The person seen just a few short years ago as a heretic and traitor could now lead the department!

RESHAPING INSTRUCTION IN LANGUAGE AND LITERATURE

The rapid growth of the IEP happily built a demand for German language classes that would not have been there otherwise and enabled the department to argue for new faculty at a time when the program had been reduced to an all-time low of two people. This meant not only the opportunity to rebuild to faculty levels of years past, but also to reshape the program to the realities of a new kind of student. When searching for new German professors, therefore, I looked for people who were not only sympathetic to the IEP, but who could bring specific strengths to the concept of a German studies major and to specialized classes for engineers, as opposed to the traditional German literature major. We were very fortunate to find a colleague, for example, who had been a physics major as an undergrad before doing his Ph.D. in German. Two other colleagues had experience and expertise related to the business world and teaching German classes with business content and another colleague had sufficient science and math background to enable creation of intermediate courses with a technical content. We became a team of five professors, all of whom value the IEP and are devoted to its students and willing to go the extra mile to support them and to ensure the success of the program. At the time of this writing, for example, two faculty are touring Germany, visiting companies, such as BMW and Volkswagen, and our partner university in Braunschweig with twenty-two freshmen and sophomore IEP students, giving them a first experience in Germany and thereby encouraging them to stay with the program and their German studies.

The early and swift successes of the IEP in German clearly caught the attention of other language faculty at URI and those interested in other language and culture areas. The faculty in French began to discuss options with me in the late 80s and were strongly supported by the College of Engineering administration where Assistant Dean Richard Vandeputte was himself a speaker of French. I personally was concerned that a French option would draw from our same pool of existing students, thereby diluting our critical mass needed to be able to offer special language sections in German for engineers. Those fears were unfounded, however, as we soon learned that developing a French IEP would expand interest in the program and ultimately increase our overall numbers. The dean of the College of Arts and Sciences was convinced that the next hire in French should be a faculty member devoted to the French IEP. Professor Lars Erickson was found at that time, a French literature Ph.D. who had studied chemistry as an undergrad and actually had worked in industry as a chemist before going to graduate school in French. Lars was attracted to URI, largely because of his enthusiasm for the IEP, and has been directing that arm of the program since then.

The next step was the Spanish IEP, which was launched with yet another grant from FIPSE and the hard work of Professor Robert Manteiga. The initial impetus for the Spanish program came from one of our IEP Board members who felt we needed to be preparing American engineering students for work in Latin and South America. We realized too at the time that Rhode Island had become a destination for many immigrants from Latin America and that a Spanish program could be a good tool for recruiting the best and the brightest of these students to engineering at URI. Students from these ethnic groups needed nurturing and guidance to understand their native language skills were an asset rather than a handicap and they could build on these skills in preparation for internships and study in both Spain and/or Latin America.

In developing these programs, IEP language faculty also refocused their research scholarship. Much of their writing and publishing is either pedagogical in nature and tied to the teaching of languages to engineers, or it is program- or curriculum-based, designed to explore and disseminate the concept of the IEP to language and engineering faculty across the country.

It is fair to say that languages have flourished at URI since the advent of the IEP. The language faculty have experienced substantial growth and prestige in the past two decades in contrast to most language programs in higher education across the country, and they have seen their department become much more central to the URI undergraduate curriculum. At the time of this writing, URI has over 125 German majors, approximately 135 French majors, a new rapidly growing program in Chinese, and a Spanish program literally bursting at the seams. The popularity of languages at URI and their good favor in the eyes of the administration may be traced in large part to the IEP and our overall notion that language learning must be rooted in many subject areas and tied to the phenomenon of internationalization, global awareness, and the need to keep the nation informed and competitive. Americans will nod politely at the idea of learning languages to broaden one's horizons, but they will not show up voluntarily in the classroom until they are convinced that it will impact their lives and strengthen their potential for career success.

Most language faculty agree that the IEP has been good for the department and for the status of languages at URI. Yet, the relationship with language colleagues has not been without its tensions and a certain amount of ambivalence among a vocal minority. There is still a residue of philosophical difference for those who see languages as the heart of the humanities and, therefore, literature-based, with a fear of committing to anything applied, as it may suggest that we are in some way in the service of, and thus subservient to, another discipline. There are also those who think that I have built an "empire" with the IEP and its two buildings across the street (see IEP housing discussion below), and have far too much leeway and authority, indeed that I am soaking up far too many resources from the university. Currently, there are a few colleagues critical of our efforts to build a Chinese program, fearing that any positions won for Chinese will mean positions lost from the traditional language programs. For me it is the ultimate irony that the only reservations at the university about building a Mandarin Chinese program have come from the language department.

DEALING WITH PROGRAM GROWTH

As the IEP grew, a major landmark in the program's history was the hiring of a full-time assistant director in support of the overall program. A national search brought Kathleen Maher to the program, a very capable and enthusiastic professional, who came to the IEP with excellent background in study abroad, fluency in Spanish, and a Masters in Latin American Studies. Kathleen brought the organizational and interpersonal skills necessary to create promotional materials, design and implement a first-class quarterly newsletter, launch and manage an annual professional conference, run a residential facility for IEP students, advise students, and so forth. She also represented the first non-faculty professional for the program, and thus marked an important step forward organizationally. Unlike faculty who could serve for a period of time, rather than teach or serve in some other way, she was devoted 100% to the administrative support of the program. The IEP was fortunate to find the right person at the right time and to be able to keep that person on board for critical programmatic steps and development. But, here again, getting the university to commit to this was not easy.

Kathleen had come first of all by means of a FIPSE dissemination grant, intended to help us continue to grow and spread the word. Having welcomed the grant as a step to help me with a program that was rapidly becoming unmanageable for one person, it was generally understood that the university would support her position at the conclusion of the grant. Not surprisingly, however, and all praise for Kathleen aside, her position would not be funded without a battle. Though I do not like to call on our Advisory Board for basic university issues, it took pressure from them to secure Kathleen's position when the grant funding expired.

Finances have been a consistent hurdle for the IEP, as has becoming fully institutionalized with adequate personnel in place to maintain the program and take care of students' needs. It is worth noting that every one of our positions dedicated to the IEP was first established with outside funding, all from grants or donations that I myself secured. Two of our German faculty were sponsored first by grants; our Housing Coordinator is funded by a grant at this point, and will be supported in the future by funds generated from the student housing payments. Our two Chinese language teaching faculty are supported by grants, as is our new Coordinator for the Chinese Language Flagship Program. I myself continue to receive summer salary support for full-time work solely from extramural funding. Thus, even after twenty-three years, we continue to build and support the IEP with marginal help from the university. This is due to its interdisciplinary nature and the fact that it is not organizationally rooted in either the College of Engineering or the College of Arts and Sciences, though its faculty are. This has been complicated by the fact that internationalization was for many years not high on the agenda of the senior leadership of the University of Rhode Island, even though the IEP has always been praised by administrators on all sides. The program did not have annual budgetary operating dollars from the university for its first twenty years! It has become clear that the program thrives on its interdisciplinary character, and yet, at the same time, this has given rise to administrative resentments, as one college felt it bore more of the financial burden than the other. In the long run, the university and the IEP have had to find a way to show that the costs of the program are borne equally by the two colleges. Even though interdisciplinarity is "in" today in academia, university organizational structures are not readily or sufficiently flexible to support its development when programs truly do cross existing disciplinary borders.

ADDING ACADEMIC EXCHANGES TO IEP

Today IEP students spend a full year abroad, first as exchange students at partner institutions and then as interns with partner companies. Given the number of IEP exchange partnerships with universities in Europe, Asia, and Latin America, it is perhaps surprising that university exchanges were not a part of our original plan. The addition of exchanges came about when the wife of President Bernd Rebe of the Technical University of Braunschweig (TU-BS) came across an article about the IEP and brought it to her husband's attention. He, in turn, asked Dr. Peter Nübold, Head of the Braunschweig Language Center, to contact me and see if there was interest in collaboration. With this first contact, a joint venture was begun between these two institutions, which has not only led to the impressive numbers of students involved, but to major research collaborations among faculty at the two schools, as well as model dual-degree programs at both the masters and doctoral levels. Peter Nübold and I have become personal friends who deal with the day-to-day issues of the exchange throughout the year, and who have nurtured the relationship between the two schools so that it impacts not only engineering students, but also faculty and students in business, the biological sciences, pharmacy, and chemistry. At the time of this writing, the two schools have exchanged over five hundred students.

It has always seemed ironic to me that I, the humanist German professor, would be a major player in bringing engineering faculty together for transnational collaboration and the exchange of engineering students. Indeed, I remember wondering if I would be taken at all seriously by engineers on either side if I suggested their students be exchanged, or whether anyone would listen if I suggested we launch a Dual Degree Masters Program, leave alone Dual Degree Doctorate. But these things have all come to pass, and the engineering faculty greatly appreciate the personal and professional connections that have been made with colleagues abroad. Indeed, this part of the journey has taken on dimensions I could scarcely imagine at the beginning of our relationship with Braunschweig.[1]

GOING INTO THE HOTEL AND RESTAURANT BUSINESS

As a result of the disintegration of campus residential fraternities, URI's centrally located Sigma Alpha Epsilon (SAE) house, right across the street from the language department, found itself vacant and without a purpose. Even though badly deteriorated, the building had a solid shell and stood on an ideal central campus location, with easy access to classroom buildings. When given the opportunity to bid for its use, and bolstered by my own one-time building contractor experience, I leapt at the chance to create a campus-centered administrative and residential facility for the program, soon to be known affectionately as the IEP House. I was convinced that students would love the idea of theme-based housing for the IEP and that they would benefit from mutual support while making their way through this rigorous curriculum. Though one might question the sanity of any faculty member wanting to create housing for students in the building where he/she works, I was convinced that a home for the IEP and its students would create a special place for increased student/faculty interaction, and it would help the program thrive.

Encouraged by the IEP Advisory Board and by the University administration, I was able to put a financial package together to launch a $700,000 renovation, creating new offices and housing for forty IEP students. After a tension-filled six-month renovation, which required a great deal of personal involvement, the IEP House was occupied by myself and thirty-six students in fall 1997. It exists, since that time, as a financially self-supporting, independent campus facility for the program and its students. The IEP manages the house, and it pays the bills by means of student rents and proceeds from special summer programs. The students govern the house themselves by means of an elected house council.

Beginning in the second year of the IEP House history, the program was able to hire Mark Schoenweiss, a highly talented chef with his own pedagogical spirit, who has endeared himself to the IEP students and the IEP community as a whole, and spoiled us with his outstanding cooking.

---

[1] I do not repeat information here that is available in previously published articles. For information about the exchange with Braunschweig and the dual degree programs at the graduate level, see: Grandin, "International Dual Degrees at the Graduate Levels," 2008.

Mark cooks for all residents, as well as for hungry faculty and other non-resident IEPers. All enjoy the food at a bargain price as well as the camaraderie of being able to share the lunch or dinner table with others committed to similar personal and professional goals. Mark takes it upon himself to make sure IEP students develop a global culinary sophistication appropriate to the careers for which they are preparing.

Given the success of the IEP House as a home for the program and the continuing downward spiral of the American fraternity system, it is perhaps not surprising that the Chi Phi House next door to the IEP House also became available. The IEP leadership soon made plans to purchase and renovate that building and incorporate it into a new two-building Center for International Engineering Education. After a $1,000,000 fund-raising campaign, the Texas Instruments House opened its doors in 2007, and its dedication on September 28 coincided with the 20th anniversary celebration of the founding of the IEP. Thanks to its donors and support from the NSF, the building created more office and meeting space, a new kitchen and dining room, a video-conferencing center, and housing for thirty-seven additional IEP students, as well as space for short-term visiting scholars. It boasts a Max Kade German Language Learning Community on the second floor, a Spanish language floor, a Sensata Technologies Living Room and the most modern of telecommunications and networking capabilities, enhancing communication with partner universities throughout the world and supporting distance learning programs. In thanks for her support in purchasing the second building, the two-building complex is now known as the Heidi Kirk Duffy Center for International Engineering Education.

MAKING URI AND THE IEP A NATIONAL LEADER

In considering why the IEP "happened" at the University of Rhode Island and not at a larger or more urban institution, it is important to note the relevance of its place as a research and teaching university vis-à-vis other schools across the nation. Being a smaller state university with a strong commitment to undergraduate education enabled us to move forward with a student-oriented, programmatic idea such as the IEP when other universities would have seen such a labor-intensive program as a distraction from traditional faculty research. This is not to say that URI does not pursue or value classic research in engineering or the humanities. It is to say, however, that its smaller size and commitment to undergraduates allowed a freedom of action that would have been more vigorously discouraged at larger and more "prestigious" institutions.

Over the years, I had gathered a network of contacts with faculty at other U.S. institutions working toward similar goals, and consulted with them about the possibility of meeting to exchange ideas, to learn from one another, and to share our collective wisdom with those who would also like to prepare their engineering students for experiences abroad. Thanks to encouragement and financial support from FIPSE, I called a group of approximately 30 such persons together in the fall of 1998 for a two-day meeting in Rhode Island, which ultimately gave birth to the Annual Colloquium on International Engineering Education. Now in its thirteenth consecutive year, the Colloquium has evolved into a regular and respected annual conference. The Colloquium is a distinctive and unusual conference in that it brings together a truly interdisciplinary group each year, representing all constituencies invested in the internationalization of engineering education. Generally, the Colloquium attracts at least 160 attendees each year and expects to remain in place for years to come as the major forum for the exchange of ideas on global engineering education.[2]

BUILDING A CHINESE IEP

Thanks to the encouragement and wisdom of both Board members and other associated business leaders, the IEP took positive steps in 2005 and 2006 to create a Chinese component of the program. It was clear to all that business and engineering were moving to the East and that no one could speak of preparing engineers globally without being able to send them to China. With seed money from Sensata Technologies and its CEO Thomas Wroe, and a soon-to-follow endowed scholarship fund from the Hasbro Corporation, both coming to us by way of URI alums, the IEP leadership moved as swiftly as possible to strengthen Mandarin language teaching at URI and to create both study abroad and professional internship opportunities for IEP students in China.

With the help of Professor Yan Ma of URI's Graduate School of Library Science and Information Studies, I was able to make contact with the leadership for educational programs at the Chinese Consulate in New York and subsequently with the Ministry of Education in Beijing. Thanks to the fact that these persons were enthusiastic about the possibility of linking language learning with engineering education in the U.S., the IEP was able to submit a proposal and receive funding from Beijing for a full-time faculty member in Chinese at URI for the initial three years of that position. Shortly thereafter, I collaborated with Professor Ma who spearheaded the drive to have URI designated as a site for one of Beijing's U.S. located Confucius Institutes, thus making URI and the IEP an important center for Chinese language and culture matters in New England.

Again with Professor Ma's help, the IEP was soon able to establish a relationship with one of China's very top engineering universities, Zhejiang University in the city of Hangzhou. We visited China together and laid the groundwork for summer-intensive programs and long-term exchanges, both of which are now in full operation. As of 2010, the IEP is in its fifth cycle for summer intensive language training in Hangzhou. 2007-2008 marked the first full-year experience for engineering students studying for the fall semester at Zhejiang University and then completing six-month corporate internships following that semester.

The Chinese IEP took a significant step forward in 2008 when it successfully competed for funding from the National Security Education Program (NSEP), enabling it to join the prestigious Chinese Language Flagship. [3] NSEP funding is intended to support attainment of the highest proficiency standards for American students studying the language, and it is commissioning the IEP to create curricular and programmatic mechanisms to encourage significant numbers of American engineering students to meet these standards. The IEP envisions significant growth for its Chinese program over the coming years as it works with local schools to create a seamless Chinese lan-

[2]For information on the Annual Colloquium, see: http://www.uri.edu/iep/colloquia.
[3]For further information, see: http://www.thelanguageflagship.org/chinese.

guage learning pathway through the high school and college years while strongly encouraging its coordination with the goals of the STEM subject areas.

MAKING IT ALL WORK AT URI

As I look back over the twenty-three year history of the IEP and my own involvement, it is clear that the program has relied on my personal commitment, my willingness to work long hours and my persistence in overcoming the unfriendly sides of academic bureaucracy. As a program that is bound to two different colleges and academic traditions and seeks the allegiances of students and faculty in new and different ways, the IEP has always been destined for challenge. I thus found myself from the very beginning in seeming conflict with people, rules, regulations, and traditions. My personal challenge has always been to find a way to get it done in the face of those who say: "No, you can't do that!"

What I have found to be lacking so often at URI is the will or mindset to make things happen when common sense indicates the action is right. When we should be asking how we can jointly bring about a good thing, there is always someone with a certain degree of authority probing diligently for some reason why we should not be able do it. I have found this to be discouraging on so many occasions and have had to accept that things will often be more complicated than necessary. The reasons for this vary, sometimes deriving from fear of change, or lack of vision, or a perceived affront to authority. Whatever the cause, the result for me has been persistence, usually yielding long-term success, but not without risk, anxiety, and, on many occasions, a definite loneliness.

To cite a specific example, I was ecstatic when my proposal to the Chinese government for the funding of a full-time language lecturer was approved. Just about everyone was happy that the Chinese would give us $40,000 a year for three years to pay for a lecturer whom they would help to select through an international search process. But, when I began the process of transferring the funds from Beijing to URI, the assistant provost declared that this was, in his opinion, a grant, and that I would have to go back to the Chinese to get overhead at the rate of 46%. The dean's office then joined the discussion, arguing that they did not have the money to pay for the lecturer's health insurance. "Patience, John! You got the attention of the Chinese Ministry of Education, and you secured URI's first full-time faculty member in Mandarin Chinese for the cost of health insurance, and now the powers that be are balking?!"

In 2002, I faced a significant fundraising challenge for our living-learning Texas Instruments House, one of two buildings where we house IEP students and our administrative offices. Our Board Chair had given us the funds to buy our second building, but we discovered soon that the renovation costs would be far higher than anticipated. We had acquired the building that we wanted to name for her, but we were short at least $500,000 to make the project work, and it was clear that there would be no state funds for this project. In short, it looked like the project was dying.

After much thought, I came up with a plan to re-channel a significant gift for an IEP endowment into a fund for house renovation. My problem was that I knew the engineering dean at that time would not want me to request that of the donor, nor would the URI Foundation like to see an endowment used as expendable funds. The latter believed that endowments were untouchable and the dean thought he could get more money from this donor for his own projects. I was really on shaky ground with this idea but felt I should contact the donor with my idea anyway. I knew how strongly he believed in the IEP and how unlikely he would be to support something else at URI.

The result was as predicted. The donor loved the idea and the dean was angry, as was the president and the URI Foundation. It is not that the president does not like the idea of a Texas Instruments House and a Heidi Kirk Duffy Center for International Engineering Education at the entrance to campus. He rightfully did not think that faculty should be messing around with donors on issues such as buildings and naming rights, which I fully understand and accept. Well, all of this came to pass and the building is now renovated and complete and the infusion from Texas Instruments encouraged other major donations of almost $500,000 for the project. Today, everybody is happy, and the donor has given even more to the IEP. I would do the same again if necessary, despite a month or so of really bad sleep. As mentioned above, academic entrepreneurialism can be lonely, even though rewarding.

A further example was the plan to create the Dual Degree Masters Program in collaboration with our German exchange partner, the Technical University of Braunschweig. This was an idea that began with one of our Braunschweig exchange students who discovered he had the preparation to do grad-level courses in civil engineering and thought maybe he could stay an extra semester or two and complete a URI Masters. After examining his record at Braunschweig and talking with some engineering colleagues, we decided to put forward the idea of a two-year dual MS program, with half the work done at URI and half the work done at the home institution. Students would complete the thesis at the host institution and otherwise satisfy all requirements of both schools and then receive degrees from URI and TU-BS. Developing this program, now universally accepted by all sides, with bragging rights for all, initially brought out multiple naysayers who raised such issues as the following:

"German students cannot apply for graduate work here because they do not have the prerequisite bachelors degree!" It took long hours to explain that the German system had no bachelors at that time and that the first degree was at the MS level. Only after long arguments, especially with the dean of the URI graduate school, could we deal with individual cases and speak in terms of equivalencies.

"How do we know they are up to our standards?" This was a cry from faculty on both sides, ultimately to be dealt with by finding grant money to support mutual faculty visits to both institutions. Once faculty knew the two institutions and each other, they were comfortable with the idea. But not before.

"ABET (Accreditation Board for Engineering and Technology) certainly will not approve this!" This was a false issue since existing requirements were being met. But tossing out ABET is a standard knee-jerk reaction among engineering naysayers, seeking reasons to be skeptical about any new idea.

"I don't want my grad assistant leaving here after one year! He/she is working for me." This is a real issue, basically only satisfied when faculty from both schools are truly collaborating with each other. When that happens, everybody wins, but it is complicated to reach that point.

"State agencies and governing boards will never accept the idea of granting dual degrees." This was an outcry from both sides, which was quieted through lengthy approval processes at both schools, and the recruitment of support from key faculty, deans and other leaders. In Braunschweig, for example, the key moment came by way of the charismatic leadership of Braunschweig President Bernd Rebe, who took a key group of faculty to lunch, gave us each a couple of glasses of wine, then sat them down in his office and said: "Gentlemen, how are we going to make this happen?"

A final example of our struggles may be seen in the doubts about the IEP in the earlier years at the presidential level. I thought President Robert Carothers would be a great ally from the beginning of his office in 1991 because he wanted change and was a strong advocate for "a new culture for learning," in which students and faculty would be encouraged to cross the disciplines and learn experientially in new modes and configurations. When taking the case of the IEP to him and asking for his support, however, he turned me down, citing it as a complex, expensive, and basically impractical model. Having been involved in building a campus in Japan for the Minnesota state college system at his last position, he seemed to be soured in terms of doing any serious international program building. But the meeting, despite the undesired outcome, did have one truly positive impact: It caused me to dig in my heels and decide to push forward regardless, determined that we could find support through grants and private sector donations, with or without the university's help.

Indeed, we were fortunate to maintain ongoing support from FIPSE (U.S. Department of Education), and when I took my plight to the IEP Advisory Board, they agreed that the IEP should "privatize" to the extent possible, and they began at that time to open their pockets, providing the funds we needed to further develop the program. The most notable commitment at that time came from Board Chair Heidi Kirk Duffy, who not only provided an annual gift for our day-to-day needs, but pledged a life insurance policy to eventually provide an endowed chair in applied German language teaching.

The commitment of the Board, of course, caught the President's eye, who began to take a closer look at what we were doing, ultimately to realize that this program was not only good for engineering and language, but also for the university as a whole. It was thus not long before the IEP began to surface in university literature and in presidential speeches illustrating the progressive power of the University of Rhode Island. Despite his eventual recognition and support, however, the IEP never enjoyed a University budget line until 2007 when it was finally embarrassed by the Board into providing an annual modest amount.

SOME LESSONS LEARNED

This brief account of my travels with URI's International Engineering Program offers some significant lessons about the challenges and fragility of international programs in the United States. Though every institution has its own character and its own set of players and therewith its own potential programs, there are nevertheless common challenges, guidelines, and principles that may be considered:

1. For any such program to succeed, there must be at least one person passionately committed to the concept who has the ability, energy, and experience to function as the driving force over the long-term. In my own case, I was fully prepared to make a professional jump from Kafka scholar to IEP director, and I was fully prepared to put unlimited amounts of time into the success of this idea. By that point, I had also gained and was gaining substantial administrative experience, serving a number of years as head of the URI German program, five years as Associate and Acting Dean of the College of Arts and Sciences, and six years as Chair of the Department of Languages. A program like the IEP will not evolve based on good ideas and short bursts of energy. It requires a driving force, sixty-hour work weeks, few vacations, and a commitment for the long run.

2. If any school is serious about internationalizing engineering education, there must be dedicated, passionate, and credible people involved at the leadership level, and, ideally, with the support and flexibility of the institution. The fact that I was already a full professor at that time simplified the process for me. Would my university have supported me through the tenure and promotion process based upon my programmatic and curricular-based publications? Perhaps, not then, but I believe it would today.

3. Passion and commitment, however, are of little value without clarity of vision as well as the organizational skills to decide where best to place one's energy. The IEP, i.e., a five-year program leading to degrees in both a language and an engineering discipline and featuring internships and study abroad, is a straight-forward and simple idea, to which one can bring energy, commitment, passion, an entrepreneurial spirit, persistence, and organizational talent. But, without each of these, the clarity of the idea can be easily muddied and lost.

4. Personality matters. An IEP director must be able to work tactfully in order to command respect across the disciplines and cultures. Because the program brings parties together who do not know each other and puts forward ideas which are often in conflict with the daily patterns and rules of the system, he or she must have patience and be able to persevere in winning over all parties to the goals of the program. Such a leader must often be prepared to move forward, even though there are many throughout the system who will say it cannot be done. I remain convinced that progress can always be made as long as the idea is sound and basically beneficial to all parties concerned. But even when those criteria are fulfilled, one still needs to be braced for the unexpected.

5. Success of any cross-disciplinary effort such as the IEP must yield clear benefits for all sides. In our case, we saw qualitative and quantitative advances for both the engineering and the language sides. Highly qualified and highly motivated students choose URI for the IEP; global companies flock to our support because of the benefits they can accrue; funding agencies and

foundations likewise come to our help because they want to back demonstrably good and proven ideas. Without clear benefits for all parties, success will be short-term, at best.

6. Depending upon one's own interpretation, the IEP has succeeded because of a certain amount of good fortune, e.g., the right people at the right place at the right time, or because of the leadership's ability to recognize the right opportunities and then take advantage of them with rigor and discipline. Be alert to opportunities of the moment and of one's own institution and do not let them pass by.

7. Engaging language faculty is necessary. Language teaching, cultural preparation, organization of the student exchange, and outreach to global companies, to international foundations, and to governmental support resources all would have been much more difficult without the help of the language faculty. It is disheartening to see well-meaning, but inexperienced engineering groups try to go this route without the cooperation of colleagues needed from other disciplines.

8. Outreach, marketing, and recruiting are absolutely essential. Because the idea of combining language and engineering study is far from the minds of high school students, teachers, guidance counselors, and even parents, the IEP depends on a rigorous program of outreach to bring the concept to the attention of potential students before the start of their college years. We might wish that the wisdom of the program were commonplace and obvious to all, but we believe the American public still is years, if not decades away from this stage.

9. Advising, mentoring, and nurturing of students in the program is likewise critical. IEP students have a challenging curriculum and, at age eighteen or so, sometimes find it difficult to focus on the outcome of the program. We, therefore, reach out to them in a variety of different ways. One key to our success is our residential learning community enabling IEP students to live together, to support each other and to have ready access to the program administration. Regular meetings, newsletters, special guest speakers from industry, panel presentations from alumni, and meetings with advanced students and with exchange students from partner schools are likewise important mechanisms for keeping the students focused and on track.

10. Global networking is key for success. For us, this means continual outreach to our partners in industry as well as our partners in foundations and governmental agencies. We depend on them for student internships and eventual job placements, but also for advice and for financial support. The IEP leadership has built relationships with literally hundreds of persons in businesses, universities, and both private and public agencies in North America, Latin America, Europe, and Asia, and it is committed to maintaining and fostering these relationships for mutual benefit. Networking means a willingness to travel regularly, but also daily networking in the office by means of telephone and e-mail.

11. In this era of tight budgets and dwindling state resources, international programs must rely on extramural financial support to do their job adequately. An IEP director must, therefore,

expect to be a fund raiser and grant writer. The bad news is that this is required, but the good news is that this is a productive area in American higher education today, with clear benefits for those who have the potential to provide financial help. Foundations, funding agencies, corporate leaders, program alums, and private individuals readily understand the importance of preparing young Americans to be savvy in the global marketplace. If a program is solid and doing its job right, it should not be extraordinarily difficult to find financial supporters.

CONCLUSION

The work is never finished. With all of it successes, the praise and awards received, it has nevertheless been a struggle to define the point at which the IEP can or could declare itself fully institutionalized, with an infrastructure ensuring a long-term future. Much of this lack of security may be attributed to the fact that I as Executive Director was, according to official organizational charts, not the IEP Director funded by a neutral or shared university line, but rather a Professor of German funded by the College of Arts and Sciences. As I have now just retired at age seventy, this problem has been resolved through the recasting of the position as a university directorship, to be funded jointly by the deans of the two colleges involved.

It speaks well of the program that it enjoys the strong and unequivocal support of the deans of both colleges, Arts and Sciences Dean Winifred Brownell and Engineering Dean Raymond Wright, who have agreed to share the main personnel costs for the program as we go forward. Both deans fully understand the complexities of building cross-college programs, but also fully appreciate the benefits of the program for their own constituencies. Of all engineering deans, Dean Wright is the most outspoken champion since the days of Hermann Viets and argues publicly and frequently that the IEP is the number one strength of the URI College of Engineering, both in terms of attracting high quality students, and in terms of national reputation among peers across the country. With the deans' strong support and participation, we have succeeded in finding and hiring a new and permanent, full-time IEP Executive Director. All parties believe that this step will mark the end of the growing pains and the full birth of the IEP as a truly cross-disciplinary International Engineering Program.

At the suggestion of URI President Robert Carothers, who was concerned that my retirement could well lead to the loss of much of the IEP's contact network across the globe, I agreed to continue my work part-time to assist Dr. Sigrid Berka, the new IEP leader, as needed, and to ensure that she take full advantage of the scores of companies, foundations, public sector contacts already in place and supportive of the program. Sigrid and I, therefore, were able to work together for the 2009-2010 academic year, before commencing my full retirement.

Has my own journey with the IEP been worthwhile? I can answer that question by saying how fortunate I have felt to go to work each morning with a sense of challenge and excitement, even at age seventy. But I can answer best by expressing pride in our students, such as those thirty currently abroad, completing their semester of study in Germany, France, and Spain, followed by their six-month professional internships. At the super-star level, I can point to John Ellwood and

Andy Marchesseault who not only completed the undergraduate IEP and then the dual degree masters in cooperation with the Technical University of Braunschweig, but who were then invited to stay on in Germany to complete their doctorates. Or I can cite the accomplishments of so many alums, who have secured fine positions with global companies such as Siemens, GE, Dow Chemical, BMW, Boeing, or who have been admitted to graduate programs at fine schools such as Princeton and MIT, or who have even gone on to law or medical school.

If I go back to my own beginnings in the field of global education and my initial experience as a twenty-year old in Germany in 1960, I realize that, imperfect as many things are, the goals I set for myself in my early years at URI have been met in many ways. I could, in a sense, be satisfied if there were just a hand-full of IEP successes, but there are in fact at least thirty IEP'ers going abroad every year, not only to Germany, but to Spain, to France, to Mexico, and to China, and we have over three hundred alums in the global workplace, almost half of whom attended a recent IEP alumni gathering associated with my retirement.

I hope this account helps illustrate the degree to which persistence, hard work, clarity of vision, patience, and tact are required for a program such as the IEP. I hope it also illustrates that remarkable progress can be made when an idea is sound and beneficial to all parties concerned. But I hope it also illustrates that life provides opportunities for the taking. At a time when I myself was almost ready to leave my profession, I was suddenly confronted with an alternative within my field, enabling me to achieve the dreams I had first formulated as a twenty-year-old. Finally, I hope it illustrates that my journey has been and continues to be extremely rewarding.

REFERENCES

Grandin, John M. "International Dual Degrees at the Graduate Levels: The University of Rhode Island and the Technische Universität Braunschweig." *Online Journal for Global Engineering Education* 3, no. 1 (2008). 332

The Language Flagship. Chinese Language Flagship. http://www.thelanguageflagship.org/chinese. Accessed August 28, 2010.

University of Rhode Island, International Engineering Program. *Colloquia on International Engineering Education*. http://www.uri.edu/iep/colloquia. Accessed August 28, 2010.

CHAPTER 16

Opened Eyes: From Moving Up to Helping Students See

Gayle G. Elliott

THE FIRST TEN YEARS

Although I did not realize it at the time, my interest in international education began when, as a senior in high school, I became one of two students from the Cincinnati area to join a New York-based student group tour to the Soviet Union. Years later, I accepted a position in the University of Cincinnati College of Engineering where I assumed responsibility for (initially part, and later all of) the International Engineering Program (IEP). When I began working with the program, it was very much in its infancy. The first group of students had not yet completed the language instruction preparing them to work abroad.

My road from senior trip to IEP, however, was not straight. I graduated from high school in the mid-1970s. Neither of my parents had gone to college. My dad went directly from high school to the Navy, where he served as a salvage diver in exotic places like Korea and Japan, and my mom went to work in an office (she did earn a degree more than twenty years later). My parents would have supported college, but unlike most families today, college was not an assumed goal. And in my senior class, while we had a few students who immediately pursued professional degrees, most of us headed toward secretarial jobs or the local hospital's two-year nursing school. I had three goals: a job, a car, and my own apartment (in that order). I took a staff position at University of Cincinnati, and other than a trip to Mexico, did not have the chance to travel internationally again until my honeymoon ten years later – at which time we spent a month in Europe. That trip included traditional locations (England, France, and Italy), and with my taste for the less common, I insisted we include Greece. The experience (and our two-income family) whetted my appetite for travel again. We began traveling regularly outside the US – mostly to Europe, Mexico, and the Caribbean, where we enjoyed scuba diving.

My work life was also progressing. I had received several promotions and was now a staff assistant at the medical center. I enjoyed my job but did not feel challenged. As a UC employee one of my benefits was tuition remission. I decided to take a class, then another and another – just for fun. I realized I was accumulating a number of credit hours and decided to apply them toward a BS degree in business (management). Toward the end, eager to complete my degree, I was taking twelve credit hours each quarter while working full time. I eventually graduated magna cum laude. (I later realized a reason education is so rewarding is because it is one of few things for which success depends almost solely on the effort an individual is willing to invest.) With a full-time job and working on a degree full-time I did not have time for much else, except during UC breaks, when we continued to travel. As graduation approached, I felt I had had enough of school, and was thinking about the future of my career. By this time, I was program coordinator for the medical staff at University Hospital. I loved the health care environment, but management jobs were often filled by clinicians. I wanted to remain at the University and felt the main campus might offer more opportunities. I accepted a lateral move, working for the associate dean in the College of Engineering. I knew it would be a step backward, and it turned out, initially, to be a much bigger step backward than I expected. I felt like I had made the biggest mistake of my career but was not sure how much of my feelings could be attributed to the fact that I was out of my comfort zone. I did not want to jump back into that zone without giving my new job a chance. At the same time, I decided to pursue a master degree in health planning and administration. It would take me two years of full-time studies plus time to write a thesis. I had combined full-time school and work once and knew I could do it again. This path would enable me to give Engineering a chance while also working toward another goal. I applied and was accepted to the graduate program. I finished my MS as planned, but during that period I had the opportunity to become involved with the International Engineering Program. By the time I finished my degree, I was fully committed to the program!

PROGRAM HISTORY

It would be difficult to continue my personal geography without first giving a brief overview of Cooperative Education at UC and the International Engineering/Co-op Program. The University of Cincinnati "piloted and perfected" Cooperative Education. The program was created in 1906 by Herman Schneider, Dean of the College of Engineering. It was part of the College at that time, and it was cautiously approved on a temporary basis by a Board of Trustees who refused to assume responsibility for its (potential) failure. Eventually an academic unit equal in status to the colleges was created to manage the Co-op Program, which handles more than 5,000 student placements annually. Today UC is recognized as the first and one of the best co-op programs in the world. Mandatory programs exist for students in two colleges: Engineering and Applied Sciences and Design, Architecture, Art and Planning. Optional programs exist in the Colleges of Business and Arts and Sciences. The program expands a four-year baccalaureate degree to five years, which enables students to complete six co-op quarters during the three middle years. Many engineering students remain with the same company for all six quarters and receive progressively more responsibility with each term, usually working as full-time engineers by the last few co-op quarters, if not sooner. The International Engineering Program was created as an option within the UC Co-op Program for motivated students who wanted to become proficient in another language and gain international experience. The IEP eventually became the International Co-op Program (ICP) as it expanded to include students from majors outside of Engineering. Students with a 3.0 GPA or higher can apply to complete a series of language and culture courses, consisting of 300 hours of classroom instruction, which makes them eligible for a six-month capstone co-op assignment abroad.

In 1991 the associate dean for undergraduate programs had just received a grant from the Department of Education (DOE) Fund for Improvement of Post Secondary Education (FIPSE) to create the International Engineering Program. The program offered two language options: German and Japanese. The intent was to create a pilot program the first year, which would be available to students in the Materials Science Engineering Department. It was also planned to begin with the (graduating) class of 1995. However, when students learned of the program, there was broader interest, and students from all engineering disciplines were accepted the first year. Students in the class of 1994 also asked to be put on an accelerated schedule, which would enable them to be included, so they became the first IEP class. They began intensive language training in the summer of 1992, worked abroad in 1993, and graduated in 1994.

Because behind the scenes work at the time the grant was developed occurred before I came to the College of Engineering, I talked with the (now retired) Director of Professional Practice for his perceptions. He mentioned concerns as well as a few hurdles encountered in the early years. To begin, in an effort to increase the probability of university support, the associate dean writing the grant met directly with the President, anticipating that he would be excited about the opportunity. He was correct, and the interest generated resulted in provostal approval to provide matching funds for the grant. While the desired result of acquiring support directly from the top was achieved, the provost at the time suggested that the associate dean should have followed the appropriate chain of command and come to him first. This was a classic example of asking forgiveness instead of permission that, due to excitement from the president, achieved the desired result. It could have had the opposite result if the president had not been immediately interested.

A variety of Japanese companies were identified and letters were sent requesting commitments to hire students in co-op positions. Sufficient positive responses were received and the grant was awarded. Also included in the grant was a mechanism to place students in Germany, i.e., a partnership with the Carl Duisberg Society (now CDS International) in New York was established. The CDS was to prepare documents necessary for students to live and work in Germany and place the students in co-op positions. This arrangement was different from UC's traditional co-op program, in which students worked with Professional Practice faculty to identify their career goals and be placed in discipline-related learning experiences. The co-op program was administered in Professional Practice while the IEP was to be managed in the College of Engineering. As a mandatory component of the undergraduate curriculum, Professional Practice was responsible for the students' co-op experience but felt they did not have full control of the integrity of the international piece of the co-op program.

There were other concerns as well. Elitism was a concern: not all students could afford to participate. Would there be a negative impact on students' co-op experience and on co-op employers? The majority of faculty did not dispute the positive effects of an international experience for engineers. They did, however, question whether that experience needed to take place in the undergraduate curriculum. And, incorporating it into an already full, very structured curriculum would be challenging. In addition to the loss of co-op time, the program required significant commitments from students to complete intensive language studies. Furthermore, some employers were concerned that they would be losing their co-op student the last two quarters when they are often expected to make the most significant contributions and, perhaps, be recruited as full-time hires.

Since there was little change to the Engineering curriculum, there was little resistance from the Engineering faculty – although many at that time questioned why an engineer would ever need to learn another language. But after the first groups of students returned from their international experiences, the program became a significant source of pride. It was positively acknowledged within the college and throughout the university. Articles were often written about the students and their experiences abroad, but some bristled. Indeed, a few who had created international opportunities for students over the years felt slighted by the attention paid to the IEP. That attention, of course, was not intended to diminish the accomplishments of others. But the IEP was the first *structured* effort to provide an international component to the engineering education. At that time, almost twenty years ago, it was truly unique!

When the IEP structure was created, the challenge was to integrate language studies— sufficient to give students a level of proficiency needed to succeed in the workplace abroad—into a very full and structured curriculum. Ultimately, co-op was the avenue of compromise on several levels. The first was language. Students who were off campus every other quarter for co-op, and who had limited space allocated for electives, could not typically develop the requisite second language proficiency through traditional courses. Thus, the decision was made to shorten part of the fourth co-op quarter and immerse students into an *intensive* language program. This required students to leave their thirteen-week co-op jobs six weeks early, missing some critical engineering practical learning. It also required employers, who often filled positions with alternating co-op students year-round, to agree to have their position unfilled for half a quarter. If an employer did not agree, the student might need to forego working for the entire quarter. In reality, only two students in more than fifteen years were told they could not be accommodated for the half quarter.

The decision to replace co-op with language was not, at that time, made because of the value it provided to the students' co-op education. It was basically viewed as the less unsatisfactory of options. There simply was no other component in the students' curricula sufficiently flexible to be removed and replaced with language studies. Although Professional Practice, was not particularly keen on the idea, it was reluctantly accepted. It was not until later that we realized the end more than justified the means. Because the learning that occurs during the final two co-op quarters is so important, the ability to effectively communicate abroad enabled students to better convey their technical capabilities. As a result, the language facilitated a higher quality professional work assignment than might be found with little or no language proficiency. Since the international co-op experience is both professional and cultural, language skills also enabled students to better adapt and become integrated into the social structure of their new environment. As stated above, the other negative impact on the co-op program was the fact that students would change employers for the last two co-op quarters. The original idea was that students would work four co-op quarters with a company in the US, after which they would work for that same company's parent or subsidiary abroad. In actuality, that model rarely worked. With thousands of co-op employers hiring in the US, and only a handful of students participating in the IEP, the students' professional interests rarely match those of the few companies willing to send them abroad. As a result, within the first group of four students in the Japanese program and six students in German, only three worked for companies that sent them abroad. Even now, more than fifteen years later, the percentage of students going overseas with their US co-op employer is quite small.

THE BEGINNING OF MY ROLE

Before the first group went overseas, the associate dean for undergraduate programs accepted a position at another university, and the person I was working for assumed responsibility for the IEP. The program coordinator remained, but the new administrator felt the IEP should have two coordinators—one for each language. The existing coordinator decided to take the Japanese Program. I thought working with the German program sounded more interesting than what I was doing, and expressed my interest. I did not immediately get rid of my other duties, and it was the first of many times that I assumed more responsibility with no immediate relief through assistance.

My responsibility with the German program initially consisted primarily of promoting the program, keeping track of the students, and ensuring that they understood and fulfilled the course requirements. I also communicated frequently and coordinated placements with the CDS, as mentioned, the organization we worked with for placement of students in Germany. In the early years, we had no input into placements other than providing the CDS with a student's resume and other pertinent documents. A few exceptions occurred when I worked with co-op employers in the US to place their co-op student at a facility in Germany.

In that first year, working directly with Japanese companies, the (four) students going to Japan were placed fairly early in the process. This was due to early development of relationships with the Japanese and the fact that companies had made commitments at the time the grant was written. The German program was much different. Despite the fact that two of the six students had jobs with their US co-op employers, two of the remaining four were still unplaced only a week or two before they were due to leave. The image of those students sitting in chairs outside the dean's office, hoping for a solution to their lack-of-a-job problem, is still clear in my mind. Although placement improved in later years, I have always wanted control of things for which I am responsible. I did not believe that giving control of such a key component of the program to an outside agency was in our best long-term interests.

The coordinator for the Japanese program left the University while the first group of students was working abroad. The wife of the associate dean, who was native Japanese, got involved in a small way with employer development but only with companies in Japan. Her efforts were not focused on connecting the US co-op experience with an international assignment. I assumed more responsibility, working with students in both the German and Japanese programs, and to a small extent working with Japanese parent or subsidiary companies in the US to create interest in the program.

As in all academic institutions, funding problems are chronic. In the early years of the IEP there were adequate resources to support the program. When the FIPSE grant expired, the University allocated general funds at an increased level to support the program. Those funds supported the salaries of faculty and staff teaching intensive language courses and administering the program. In addition, they provided travel grants to participating students. We were also able to host small promotional activities designed to recruit and retain students, gain publicity for the program, and acknowledge our many supporters. Often included were quarterly informal meetings where students were able to socialize over pizza. In addition, departure and return celebrations were held to thank supporters for their roles in the program, thus enabling students to see the hard work involved in making their international co-op experiences possible. As my time became more valuable and funds became scarcer, many of those extras offered in the early years have been foregone.

Over the years my role continued to evolve and I willingly took on more responsibility. Eventually, although the associate dean was ultimately responsible for the program, his other activities took precedence, and I knew more about the inner workings of the IEP than he.

STRENGTHENING THE CONNECTION BETWEEN THE IEP AND CO-OP

I began working more closely with Professional Practice, which was still leery of the integrity of the program, particularly the quality of the jobs. I had met the assistant director on a few occasions and we developed a friendly working relationship. He had responsibility for the mechanical engineering co-op program and we did some job development together, targeting companies in the US with operations in Germany and Japan. For the first time, I began to tie our two areas together. I also made decisions that supported the needs of their employers rather than permitting situations of conflict that could damage their program and employer relations. Through my involvement with committees and activities on campus, I became friendly with another Professional Practice faculty member who worked with electrical engineering students.

I began to see the significant differences they saw between the IEP and the co-op program. On the surface they looked the same, but there were salient differences. Since students co-oping in the US often remained with the same company for all six quarters, their responsibilities increased each quarter, either in *breadth* by exposure to various areas within the company, or in *depth* as they became more expert in their area. The CDS, although an excellent resource, did not have in-depth knowledge of the students' professional expertise and goals. This is an area where the difference between co-op and internship became quite obvious. The students had excellent cultural experiences abroad, but the jobs were often not a good match for them professionally. And, unlike jobs in the US to which students apply, go through an interview process, and make a decision vis-à-vis acceptance, the jobs in Germany were announced by fax stating, "John Smith has been placed at *XYZ Company* at a salary of DM *XX* (currency at that time)." Another issue was that engineering students meet with their co-op advisor following the first five co-op assignments, but not following the last. A senior survey response about their experiences replaces the one-on-one meeting with their faculty advisor after the final co-op placement. Since international assignments took place during the last two co-op assignments, none of the knowledge acquired overseas was being shared with any member of the University.

I began making subtle changes, trying to create a structure for the IEP similar to the UC co-op program. I began having more in-depth discussions with students about their goals. I also started meeting with them after their return in order to assess the learning that occurred and get feedback regarding their experience. Further, I worked more closely with the CDS and employers to create a better match between students' professional interests and the jobs overseas.

In 1998, the associate dean left administration and returned to a faculty role, and I was re-porting directly to the dean of engineering. Since 1994 I had been the College of Engineering representative on the All-University International Planning Council for the President's Global Initiative. Our Committee had just completed a search for the first UC position dedicated entirely to international programs. An associate provost and director was hired for the newly-formed Institute for Global Studies and Affairs (IGSA). The IEP had attracted quite a bit of attention by this time. Other co-op colleges were interested in having a similar program and were making their interest known to the university administration. A fellow committee member, also a member of the faculty, suggested that I was in a unique position to define my own position within the program. She suggested I consider whether I wanted to pursue a faculty or an administrative role. While at one time I would have had no interest in a faculty position, working with the co-op faculty had opened my eyes to a non-traditional faculty role where most of the teaching took place outside the classroom – one-on-one with the students. I was interested in that. And I was definitely interested in taking a leadership role and expanding my responsibilities to provide international co-op opportunities to students outside of engineering. I initiated conversations and a series of meetings with the dean of engineering, the provost, and the associate provosts for Global Studies and Affairs and Professional Practice.

THE MOVE TO GLOBAL STUDIES AND AFFAIRS

The decision was eventually made to move (what would become) the International Co-op Program (ICP) to the Institute for Global Studies and Affairs (IGSA). The Division of Professional Practice was administering a well-established basic co-op program. Both associate provosts agreed that if the ICP developed beyond one college and became an established program with continued university support, it might be beneficial to move it to Professional Practice, but during the developmental phase, it would be more appropriately managed in a department focused specifically on international programs. The Dean of Engineering had only one requirement – the level of service to Engineering students must remain the same. Despite competing priorities and budget cuts, I strive to this day to fulfill that commitment.

In spring of 1998, I and the IEP (with budget) moved to the Institute for Global Studies and Affairs. There was discussion about giving me a faculty appointment, but this would create logistical difficulties since IGSA was not an academic unit. I would need a dual appointment, reporting to two associate provost/directors. I realized the need to focus on the needs of the program instead of my own career goals at that time. I accepted the position of director of International Co-op Programs. I began traveling annually to Germany and Japan to meet with employers who had hired our students as well as others with whom I was able to generate interest. Although employers rarely had negative feedback about the program, I began to better understand their needs and why they participated. I developed relationships with employers that would serve me well in future years and provide continued opportunities for students on an annual basis. Equally important, I instituted a requirement that all students meet with me on their return from abroad. These changes enabled me to capture a wealth of information.

I learned what was important to our international employers and could better prepare students to fulfill their needs. For example, employers in Japan enjoyed being able to expose their employees to American and English-speaking students. For many of them, ours were the first Americans with whom they had ever worked. They benefited from different perspectives and a chance to practice and improve their English skills. They also appreciated the students' understanding of and appreciation for Japanese language and culture. The students in Japan often received a significant amount of responsibility and often made presentations, in Japanese, to their colleagues. Employers in Germany had a long history of having German and foreign students in the workplace. They liked the diversity provided by the Americans and valued the technical skills UC students acquired during their first year of co-op experience. They were surprised that the UC students could take responsibility for their own projects and become valuable members of work teams. In both countries, I found that once companies had experience with our program, they usually continued to participate.

CHALLENGES IN THE IGSA CONTEXT

The move to IGSA provided a welcome internationalized environment for the program. The office was devoted solely to international experiences of various types and to promoting the goal of attracting more students to incorporate an international experience into their education. It was also somewhat of a mismatch. In addition to me, the department consisted of four others: the associate provost/director, an assistant director, a manager for study abroad, and an administrative assistant. In varying degrees, the role of each was to serve as a resource to faculty developing international programs for their students. Instead of serving as a resource to others, however, I was responsible for developing and managing all aspects of the International Co-op Program. This included recruiting and retaining students, hiring language faculty, preparing students to go abroad, developing jobs and maintaining relationships with employers, assessing the students' learning and program effectiveness after students returned, and making improvements to the program when necessary. The ICP had its own budget, so I was responsible for that as well. I even ventured into fund-raising, receiving more than $60,000 from alumni and other donors to support the program. I was the only member of the department whose responsibilities were not in a resource role but that of program manager and educator. This created some dynamics that at times were difficult to reconcile. I was often perceived as having a bias for my own program – which, of course, would be expected.

There were also some conflicts in philosophy and expectations between work abroad and study abroad. To study abroad for a semester or year, students usually expect to spend $5,000 - $20,000 or more, depending on the program. The target population for the ICP was different: co-op students (usually engineers) who came to UC with the expectation that co-op earnings could substantially reduce the cost of their education. An engineering student who does *not* participate in the ICP can expect to earn about $40,000 through their six quarters of co-op experience. A student who participates in the ICP reduces that income significantly. Due to heavy subsidization from the ICP budget, the student's cost for the summer intensive language program was only three credit hours of tuition, instead of eighteen. The more significant cost, though, is the eight to ten weeks of co-op earning opportunities forgone and replaced by the intensive language programs. In addition, salaries paid abroad were about one-third ($1000 vs. $3000/month) of what a student could earn in the US. As a result, the net cost to students participating in the ICP can be approximately $10,000 - $15,000.

I acknowledged the need to make the program sustainable, but I also believed it was important to understand the needs and expectations of our target population: students who calculated costs differently than would potential study abroad students. This often put me at odds with my colleagues, who felt that engineering students unfairly benefitted by the opportunity to be paid during their international experience. I was under some pressure to significantly increase the students' cost for the language program, regardless of how it might affect participation. I felt the goal should be to break even, not make a profit, unless we could do so without harming the program.

My rationale was simple. Although the ICP is a valuable, personal and professional life-changing experience, the rewards are not completely known until students actually go abroad. At the time the decision is made to participate, the financial cost is explicit while the benefits are not. To create a program that would attract students, it was important to acknowledge and accept that fact. Returning students, without exception, indicated that they would choose to participate again, but that was after the experience and the recognized value it had provided.

In an effort to reach a compromise in my department, I decided to determine whether we could increase tuition for the summer course with minimal impact on participation. I sent a questionnaire to students describing the cost of the intensive language courses and the impact of university budget cuts on the program. I asked if they would choose to participate in the ICP if the cost of the summer course increased to six credit hours, nine credit hours, or twelve credit hours. The group of students being polled were those who had just returned from an international experience, so they already knew the value they had gained from the program. (I was leery about asking those about to participate, for concern they might assume an imminent tuition increase and drop out of the program.) Nearly 100 percent of the students responded that if the tuition increased to six credit hours, they would find a way to pay the higher cost and remain in the program. With an anticipated increase to nine credit hours, sixty percent of the students said they would *not* participate. And with an anticipated increase to twelve credit hours, 100 percent of the students said they would *not* participate. Subsequently, we increased the tuition cost for the summer class to six credit hours with minimal impact on participation. At that time, I also negotiated an agreement with the university so that the program would receive tuition dollars paid by the students to cover the cost of faculty teaching the courses. With six credit hours of tuition, we were not making money, but we were breaking even.

Within a few years of my move to IGSA, my responsibilities with the program had increased to the point where I desperately needed administrative assistance. The associate provost supported creation of a twenty-five-hour per week part-time position. A skilled person was hired to provide assistance in coordinating numerous aspects of the program. Unfortunately, it is difficult to retain qualified staff in a part-time position. A few years later that employee resigned for another opportunity, and I embarked on a variety of temporary solutions to the ICP's need for administrative support. These included an ICP student, who worked part-time her senior year; additional hours for a part-time German Department staff assistant; and eventually, another part-time staff member designated to support the ICP. These solutions were separated by periods of time with no support, making it difficult to maintain consistency.

STRENGTHENING AND GROWING THE ICP

During this time I revised a seminar designed for students preparing to go abroad and developed it into a course to be offered through the Division of Professional Practice. (It was not possible to offer the course through IGSA, which was not an academic unit.) The course was designed to give students an overview of the program requirements, to expose them to the languages and cultures offered, and to introduce them to the challenges and opportunities involved in living and working abroad. The goal is to create realistic expectations in order to promote a successful experience abroad. The course meets weekly over a ten-week academic quarter. I prepared a course description, submitted it, and received approval from the Professional Practice Curriculum Committee to teach it as part of their course offerings.

I also became responsible for hiring language professors. We have a strong German Department, and the University was beginning to increase its commitment to Japanese. We utilized adjunct faculty from both departments to teach our courses. Due to the size of the German group, the summer intensive German courses were taught by an experienced adjunct faculty member assisted by two advanced level graduate students. The German Department had been instrumental in creating the program at the time the grant was received. We learned that graduate assistants with experience teaching intensive German to engineers had increased marketability at graduation, so it was a mutually beneficial situation. Although the ICP hiring decisions were mine, I relied on the expertise of language faculty when selecting those to teach for us. And, when a search took place for new language faculty, I was often included as a member of the search committee. The opportunity for an adjunct to earn additional pay by teaching for ICP in the summer was an incentive to recruit strong candidates to adjunct positions. In addition, we recognized it would be easier to maintain consistency and high quality instruction if we utilized faculty from within the University rather than searching outside annually to fill ICP needs.

One of my strengths is the ability to work with a diverse group and create and maintain strong relationships. Universities usually consist of a group of colleges operating completely independently of each other. Working across disciplines is rare because each group has different priorities, but programs like the ICP depend on it. Necessary ties can become strained when one or more groups feel they are no longer getting what they want out of the relationship. We experienced some challenges, as funding became scarcer and the language departments expected commitments we could not make, but that can be expected in any situation where two groups have differing primary goals. For the most part, we worked better together than usual because of the strong commitment of each group and because the positive aspects to each outweighed the negative.

CHALLENGES IN BUILDING A SPANISH PROGRAM

While expanding the program to include students from other co-op colleges, we also decided to create a third language program – Spanish. Some of the biggest challenges have come through creation of that program. As is typical in academic environments, there was no increase to the budget to create this new language program. We attempted to keep the cost low using an adjunct faculty and graduate students, but once again I faced the issue of charging full tuition for the summer program to "rich engineering students." The faculty teaching did an excellent job, giving more to the program than simply their classroom time. Unlike the German and Japanese programs, however, in which students had minimal to no previous language skill, students in the Spanish program all had some level of Spanish proficiency—some with excellent skills. This made structuring the course more difficult. It also made it difficult to justify the additional expense – particularly when faced with additional budget cuts. After three years of struggling to pay for a third language program on campus, we decided we could no longer afford to offer intensive Spanish.

I began investigating options and learned that the Spanish Department offered a six-week study abroad program at the Monterrey Institute of Technology and Higher Education's **(ITESM)** branch campus in Querétaro, Mexico. Students accompany a UC faculty member to Querétaro, where they take classes at ITESM and live with a host family. The program was an excellent match for our needs. Students had language immersion as well as a true cultural experience while residing with a host family. The prerequisite required for the program (one year of college-level Spanish) was easily met by students who often had four years of high school Spanish, followed by a quarter or two of college-level Spanish. The fact that all students did not work in Mexico did not create much of a problem. There were some minor language adjustments when they took their Mexican Spanish to Chile, but they adapted quickly, in part due to the cultural adjustment they had experienced the previous year in Mexico.

Jobs and salaries in the Spanish program presented another challenge that was not so easily resolved. Employers were unfamiliar with the concept of co-op or paid internships. Most were unwilling to pay students even the minimum salary needed to live in the local economy. In the second year of the Spanish program, I placed four students in Santiago, Chile. One student earned the equivalent of about $600 per month. This was sufficient for him to pay for rent, food, and local transportation, with a little left over for entertainment. Two of the others earned $200/month, and the fourth student earned only $100/month. From experience, I knew that new employers overseas were unaware, prior to working with a student, of the level at which an experienced co-op student could perform. I hoped hiring their first student would enable them to experience the value. I decided to invest in the program by subsidizing students' salaries. Consequently, the three lower-paid students received a scholarship stipend to bring their salaries to $500/month. I believed this investment would pay off as employers became willing to offer higher salaries in the future. My expectation proved wrong. Despite what the students were able to contribute, most companies were simply unwilling to pay a student anything close to what their permanent employees received. In order to survive, I realized that the Spanish program would need to rely heavily on relationships with our co-op employers in the US. This creates the added difficulty of identifying the right person within an organization who has the interest and influence to make placement abroad possible, and it is often not the same human resources person that we work with for US co-op placements. It also relies on matching a very small group of students with a much smaller than usual pool of employers. Our Spanish program is still small, but most of the placements are now made through co-op employers on the US side, or through personal and professional contacts of the students. Through our US co-op employers, students have been placed in Mexico, Chile, and Costa Rica.

The Spanish program is now facing another hurdle. The dates of the ITESM program changed, and the new dates conflict with the end of UC's spring quarter. This new challenge, however, may open the door for more options within that program. I hope to identify a few Spanish programs, ranging in length from six to twelve weeks, in various parts of the Spanish-speaking world. Students could then select a program based on where they planned to have an international co-op experience. Through this model, students would be better prepared both linguistically and culturally for the international co-op experience. UC also made the decision to change from a quarter to semester system in 2012. When that change takes place, our dates should match with the ITESM calendar, and students could again participate in that program.

LIVING ABROAD

There were many times when I felt Professional Practice, rather than IGSA, would have been the more appropriate administrative unit for the ICP. The program would have benefited from the systems in place to support co-op, and I would have had opportunities for professional growth and interaction with others in the co-op field. But by mid-2003 the ICP and I had been housed in IGSA for five years. During this time, my husband, an economics professor at UC, had created a study abroad program to Japan, through which he developed a relationship with *Tokyo Kazei Daigaku* (Tokyo Economics University). He was invited to take a sabbatical there and received approval from UC to do so. A sabbatical abroad had been a dream of ours for years, and I decided to join him. Previous conversations with the director indicated a leave or other arrangements for this experience would *not* be supported. I planned to tell him in the fall that I would resign my position effective 1 January 2004. As so often happens in life, fate took control and created another opportunity. During that summer the director notified the University that he had accepted another position and would leave UC at the end of December. The associate director was appointed interim director. Since she would assume responsibility for the department right as I was leaving, I did not wait long to tell her of my plans to resign in January. She asked me to reconsider and suggested we might work out an arrangement that would permit me to go to Japan. I was excited by the possibility and agreed to entertain the idea. As a result, I was able to keep my job and spend almost eight months in Japan. I worked part time from there and returned to the US on three occasions for one week.

My time living abroad was extremely satisfying. By then I had traveled to Japan five times for two-to-three-week periods each. I had a good understanding of the culture and knew my way around. But now I was living there for an extended period. We had an apartment in a university complex that was large by Japanese standards but small by ours. We had a Tokyo address but actually lived in Kodaira, a suburb outside Kokubunji, twenty minutes west of the center of Tokyo. This was my opportunity to experience conditions similar to those I had been creating for my students. After years of traveling to Japan, I did not experience the traditional *culture shock*, but did experience a series of small cultural *adaptations*. I had used public transportation whenever possible on my trips to Japan. It helped me become familiar with my surroundings more quickly and made me feel self-sufficient. I was very familiar with the crowded trains, but after a month of using them regularly, I longed for the comfort of my car – away from the crowds. But soon I adapted to the crowds and accepted the trains for what they are – a crowded but wonderful way to travel without the expense and hassle of driving.

Living for an extended period in a foreign country is very different from visiting. I had the opportunity to participate in more daily activities, to really experience the heartbeat of the culture, of a neighborhood. Small things like being served tea at a hardware store while waiting for a key to be duplicated or separating and setting out one's trash *every day* define the cultural differences in small but meaningful ways. It took weeks after returning home before I comfortably, without thinking, threw all our trash in the same trash can. Even today I miss bells ringing the Edelweiss melody from somewhere in our neighborhood each evening at 7:00 pm. The need to develop patience in dealing with cultural differences was reinforced by my experience in Japan, though I long ago felt it sliding away as I once again became immersed in the fast-paced, chaotic American culture.

RETURNING HOME

While I was living in Tokyo, things were changing at UC. After years of discussions about whether the ICP fit better in Professional Practice, the director's departure provided the opportunity for change. IGSA's interim director understood the benefits a move would give the program and helped convince the University administration it was a good move. Further facilitating the change was a new associate provost who had assumed directorship of Professional Practice several years earlier. He was international himself, having come to UC through an exchange program with his university in Finland. He was interested in accepting the ICP into his unit. This was good news, but my status was somewhat uncertain. I had committed to remain with the program through the end of my time in Japan. When I returned I would work with my new director to help determine the future direction of the ICP.

My strengths and professional interests are in international job development and working directly with students to prepare them for a successful international co-op assignment. These were similar to the primary responsibilities of Professional Practice faculty placing co-op students in the US. An administrative appointment in this academic unit would give me managerial control of the program, but the job development and student advising aspects would likely move to the faculty. The next academic year was beginning in a matter of weeks. After discussions with my new director, it was determined I would remain as a visiting faculty member for one year, responsible for international placements of students in the ICP and for all aspects of co-op placement for a third of mechanical engineering. Since I also retained administrative management of the ICP, this increased my workload by fifty percent. With high expectations for myself, I was confident I could handle the load. The part-time ICP assistant had accepted another position while I was in Japan, so that position was vacant again. But Professional Practice had a fairly large support staff, and I was confident I could get the help I needed. Despite my increased workload, I believed that systems were in place to support the ICP. One year later, after a national search, I was hired as a tenure-track faculty member with the same responsibilities, and four years after that I was promoted to associate professor.

NEW CHALLENGES

My new position brought challenges, but not for the reasons you might expect. For years I had engaged in professional development activities traditionally undertaken by faculty, so continuing to do so was not an issue. Working with the mechanical engineering placements in the US also came naturally, and it was rewarding to work more closely with a large group of 100 students over a five-year period. However, like in universities and industry everywhere, the budget cuts continued. Within a year, three of the nine support staff positions were abolished and the ICP was forced to rely on sporadic and inconsistent support from several different sources. It was difficult to handle the myriad of ICP administrative responsibilities in addition to my new faculty responsibilities. I prided myself on my ability to be efficient and productive but often felt I was moving in so many different directions it was difficult to keep everything straight. For a long time pieces of the ICP support staff functions were periodically given to others, but there was no formalized structure for program support, and the lack of consistency inhibited my ability to create systems to facilitate smooth operation of the program. This began to change a year ago when one of the staff assistants expressed interest in the ICP. She assumed responsibility for several key functions of the program, and, due to her strong capabilities, has provided the consistency we so desperately needed.

In addition to staffing, international programs of any kind bring a variety of challenges, and international co-op is no exception. A few of the most problematic, and descriptions of how I dealt with them, are listed below.

Fluctuations in numbers have always been an issue. The fluctuations occur to a lesser degree in total numbers and to a large degree in participation by major. I may consistently have fifteen to twenty-five students annually in the German program, but the jobs are discipline-specific, and there is little consistency of numbers by discipline. For example, after years without an aerospace engineer in the German program, in 2004, four aerospace students participated. It was a particularly difficult time to place students from that major in Germany. After utilizing all my contacts and even creating one more, I still had one less job than I had students. That student worked in a mechanical engineering job, which was somewhat related to his major, but without the aeronautic-specific focus. While the student was lucky to have a job, he felt slighted at not getting an aero-specific job. The problem is that we often become a victim of our own success. I tell students that jobs are not guaranteed, but with a 100 percent placement rate (until 2009, that is), it is difficult to create the expectation that such may not always be the case. After that year of four students, I have had zero or one student studying aerospace in any given year. Several of the jobs developed for the group of four will likely not still be there when I need them again in the future.

In a similar situation in Japan, I contacted potential new employers to create new positions when participation jumped from zero to three chemical engineering students. Two of the chemical companies I approached were interested, but they wanted accounting or finance students. I was able to create other positions for the chemical engineers, but tried (without success) for several years to attract business students to the program and those jobs. Eventually the employers gave up and looked elsewhere to fill their needs. When I finally did have an accounting student in the program, I had to start all over again to create a new job. The job development then becomes a balancing act in which I attempt to determine in the spring what my needs will be for the following spring, and then do what is necessary to create jobs for those students.

In 2009 the ICP experienced the worst case to date of such situations. There was only a small increase in the German program, but two students were from majors I had never placed before. In addition, after making an effort to increase participation in the Japanese program, I succeeded with a significant increase from four students to eleven. In spring 2008 I felt I did everything right to ensure sufficient positions were available for the group. I returned from Japan feeling good about my job development accomplishments and confident I had a match between student demand for jobs and employers willing to accept them the following spring. The first bad news came when a company, which had been very excited about one of the students, failed to get human resources approval to "accept an American intern in 2009." The second occurred a few weeks later, when the US financial market began crashing, and a company interested in the business student said they could not accept a student due to financial concerns. Two students then got placed, but three more companies, with which I have worked for many years and expected a positive response, notified me they also could not accept a student in 2009. The problems continued and in the end, despite an almost unbelievable effort on my part and from others who helped, I was faced, for the first time in history, with three unplaced students. The image of those two students sitting in the seats fifteen years ago flashed before my eyes. I knew that failure was not an option and I could not imagine having to tell students who invested hundreds of hours learning Japanese that there was nothing I could do.

While considering possible plans A, B, and C, I discussed the issue with our director who came up with an excellent idea. With potential job options exhausted, he suggested providing the unplaced students with scholarship money to study abroad in Japan. I researched Japanese language school options and identified one with flexible dates that would meet our needs. The cost was too high to support a six-month experience, but we could support each of the students for one quarter. This cut their time in Japan in half, and it would not be in a professional environment, but at least they would have an extended living experience abroad. The students seemed relieved that we finally had something to offer and understood the difficulty we had faced. One of the three actually did receive an offer during the study abroad program and was able to follow that with a three-month work experience in Japan. While I would never choose to go through this experience again, in some ways it was beneficial. Short of having a student injured abroad, my biggest fear was unplaced students, and I spent many sleepless nights in difficult placement times. Now the worst has happened and we survived with a solution. To turn a negative to a positive, it also may enable me to create more realistic expectations among students now that we long longer have a 100 percent placement rate!

SO WHY DO WE DO IT?

When approached to participate in this workshop and book project, we were asked to write our personal geographies in an effort to help others understand "what led a bunch of crazy people to devote so much of their time and energy" to this purpose. For me, and I am sure most others in these positions, it is the ability to see a potential opportunity and the desire to make the most of it. Success comes from the fact that we refuse to accept the words "can't," "won't," or "don't." Too many times an employer's first response has been, "We don't send students overseas." Some may not understand how our response in return could be anything other than, "Why not?" followed by an explanation of why they should. The people who thrive in these jobs are those who find pathways around obstacles rather than stopping.

I am not creative in an artistic sense but have always been good at creating something from nothing and improving a process that needs help. I (sometimes foolishly) believe I can control everything and ultimately achieve success. Most people want to feel ownership for their work, but I cannot imagine a job without it. The opportunity to take ownership of the German IEP attracted me to the program more than fifteen years ago. The students are what keep me here – even through difficult times. I believe most of us feel the same seeing the difference we make in the lives of our students. As a teaching professional, my contribution to their educational and career development is my work life *raison d'être*. Unlike faculty in a primarily classroom/research environment I, as a professional practice faculty member, have the unique opportunity to know my students well. Instead of a quarter or two, I see them grow over a five-year period. They come to my office as freshmen, many from small rural communities away from home for the first time. Before flying halfway across the world to spend six months in a foreign country, some of my students have never been on a plane nor traveled outside the Ohio-Kentucky-Indiana tri-state area. I can make a significant impact on their lives. I derive satisfaction from the knowledge that I play a part in introducing students to important international experiences that often shape their personal and professional direction. A former teacher opened my eyes to the excitement of another culture and made me realize the impact an enthusiastic faculty member can have on a student. Feedback from my students indicates that I generate that same excitement in my class—and that class is just the beginning. I continue to work with those students for several years, culminating in a successful international experience. Although I can site dozens of examples, I want to include only a few.

After sending an email notifying one of my students of her job offer, I received the following response, which was so exuberant that I have kept it as a reminder for more than ten years: "Thank you!!!! Oh my Goodness—sounds exactly like what I wanted!!! Thank you! Thank you!!" It is a great feeling to create that type of excitement for a student.

The next involved a male, who was somewhat immature. I had some concern about sending him abroad, but he completed the coursework and had a job offer, so off he went to Germany. His mother called me a few months after he left expressing pleasure at the new maturity she saw in him when she visited. On his return I put him in contact with a younger student from his major who had questions about the IEP. He copied me on his message to the younger student, part of which said, "No one event has changed my life as much as the IEP experience. I now fear no map, subway, adventure, entrée, beverage, or autobahn. My tolerance is now my strongest trait."

And finally, more recently, two students submitted their personal experiences about working with me in documents for my tenure dossier. The first, a young man, wrote, "I make no exaggeration when I say that participation in the ICP was the single most life-changing experience, among many wonderful experiences at UC. Gayle may be unaware of this, but she is among the ten most influential people in my life." That student went on to graduate school and is currently working with a company in Japan.

The other, a young woman, came to my office as a student to inform me she planned to drop out of the ICP. She had already completed most of the German course requirements and I was working on her placement in Berlin. She was the only female in the group that year and the only student whose job would place her in the former East Germany. She exhibited anxiety about the program, the language, and about living far from the other UC students. We talked and I expressed support for her decision, but encouraged her to further explore her feelings. She remained skeptical as she left but agreed to think more and discuss it with her parents. After doing so she decided to remain in the program. She had an outstanding experience in Germany and when she return told me the other students in the program spent a lot of time with each other but that her success was "so much sweeter because (she) did it alone." She accepted a full-time job in Germany when she graduated and now resides there with her husband. In a letter she stated,

Gayle played a very important role in my educational career that has resonated in my personal and professional life. As a student it is very important to have professors who enable them to pursue and achieve their dreams. It is an impressionable time and to be so lucky as to have a professor who is capable of identifying and meeting their students' needs, and willing to be personally vested in their success is a valuable resource for any program at the University of Cincinnati.

Of course, such feedback is satisfying. But the real satisfaction lies in the knowledge that *I can do it again* for another student! ICP students are also extremely loyal to the program and want to do what they can to provide other students with the opportunity they had. In recent years we have placed more than twenty-five students abroad due to the efforts of our alumni, who hire ICP students to work for them, or influence others in their companies to do so. It is wonderful for me to work with an employer, previously a student, to place another student abroad.

THE FUTURE

The next challenge and opportunity will be in accommodating the ICP into a new structure, as UC plans to convert from quarters to semesters in 2012. Employers have often asked if students could stay longer than the current six-month, two-quarter co-op period. The semester structure will enable us to combine two co-op semesters, thereby meeting the need of our employers for an eight-month extended assignment. It will also give students a longer and therefore more culturally rewarding experience as well as the opportunity for more in-depth projects abroad. Semesters may also provide an opportunity to more easily create a one-year international experience, combining study abroad with international co-op.

The International Co-op program is strong and evolving and challenging. As the Japanese say, "Success is the mother of failure." We cannot take things for granted, and my goal is to continuously improve and expand this program.

CHAPTER 17

What is Engineering for? A Search for Engineering beyond Militarism and Free-markets

Juan Lucena

INTRODUCTION

This chapter describes my search for answers to the key question, what is engineering for? This search began at a key moment in the 20th century when the Cold War was coming to an end, and my struggles to understand the implications of this event for engineering in the U.S. and my home country of Colombia intensified. This search took me to a key policy site where academic and corporate activists were redefining engineering for a post Cold-War world. I decided to theorize about these events and their implications for engineering education while pursuing a Ph.D. in Science and Technology Studies (STS) where I began finding answers to the question of what is engineering for in countries other than the U.S. as I co-taught the course Engineering Cultures. During my first faculty appointment, I struggled to redefine this course for new audiences and discovered the challenge of co-optation as I developed key networks of friends among executives and engineers working for U.S. corporations. In my second faculty appointment, I had to redefine the course one more time for new audiences and interests, and I found that my search for answers to the key question created new struggles in dealing with interdisciplinarity and effective pedagogies. Most recently, my search is making me consider alternatives to traditional purposes of engineering and develop possibilities for engineering for humanitarian, community development, and social justice endeavors. The chapter concludes with the 'problem of scale' and a brief analysis of recurring challenges to my efforts to further disseminate my curricular and pedagogical efforts.

FOR COLD WAR MILITARISM? FOR FREE MARKETS?

My search began with the decision to study engineering at Rensselaer Polytechnic Institute (RPI) where I began asking and finding answers to the question, what is engineering for? In the mid 1980s, my father, a lawyer, career politician, and congressman for many years, became preoccupied with the political and ideological shifts taking place in the USSR before the end of the Cold War because of their potential impact on the geopolitical configuration of the world, including Latin America and Colombia. Through my weekly long-distance debates with him, I began to think about the connections between engineering and the Cold War. Perhaps the content of my engineering curricula would not be the same in a post- Cold-War world. Would my professor in Vehicular Dynamics and Control be comparing Soviet MIGs and U.S. F-16s if the USSR were to fall apart? Would professors be giving us examples and homework that included the trajectories of ICBMs or analysis of the space shuttle's main engine? Would we be taking courses like Numerical Methods and Composite Materials, inspired heavily by Cold-War military technological development? Would students continue doing senior design projects in labs sponsored by the U.S. Department of Defense? If not, how would the examples, content, and labs be different? How would different examples, courses, and sponsors change the relationship among faculty, students and engineering knowledge? Would students who dreamed of designing fighter jets still be drawn to engineering? Or would the pool of engineering students be different in a post-Cold-War world? In short, I began wondering about the relationship between engineering and students' knowledge and desires in a geopolitical context. While taking Senior Aircraft Design, I wondered about the relation between the problems that I was solving (e.g., stability and control of an experimental sailplane to be built with new composite materials) and the organizations that sponsored the labs where I was working (e.g., Boeing, Alcoa, Pentagon). A year later while taking the class *Politics of Technological Design*, I began to understand this relationship and to realize that engineering always has political dimensions. Recognizing my own ignorance in a class paper, I wrote,

> [I]t is worrisome to realize that students of engineering [myself included] are not aware of the forces of production in the way they should be. After reading David Noble's *Forces of Production*, one can clearly see how the production of technology is influenced so drastically by few very strong forces within our society…We see today an enormous increase in enrollment in electrical and aeronautical engineering…yet these people, far from recognizing the forces of production, only recognize the supply and demand in the job market.[1]

It would become increasingly clear to me that the answer to the question, "what is engineering for?" is often shaped by power relations among those funding engineering research and those involved in educating engineers (i.e., university administrators and faculty). The voices of students appeared mostly silent in shaping the answer to this question.

As a student from Colombia, I also realized I had a different relationship to engineering than my U.S. peers. It seemed that neither faculty nor my U.S. peers were questioning the possible impact of the end of the Cold War on engineering. This was a preoccupation of a student from a "Third World" country influenced by a politician father who was trying to figure out how the future configuration of world politics would impact the country he was helping run. So my preoccupations about engineering were of a different sort. Although I did not know it then, my father was influencing me to think

---

[1] Lucena, Militarization of Technology, 1987.

about what kind of engineering would bring the most progress to Colombia in a post Cold-War world. During his early incursions into public service, which began in the mid 1960s, my father lived firsthand the rise of the ideologies of development and modernization in Colombia. But in the 1960s, he was not interested in technology or engineers. In his mind, development happened via economic policies and progressive politics that stimulated economic growth and protected nascent national industries. These were to be implemented by lawyers and economists serving as public functionaries, not engineers. [2] Although development and technology came together, engineers were invisible to him. According to tradition in my paternal family, the only acceptable professional pathways to success for family men were law and medicine, not engineering. Three generations of Lucena lawyers and doctors had brought much prestige, power, and wealth to the family and, according to some, "progress" to Colombia. By the mid 1980s, I was breaking away from family tradition by becoming the first engineer. Perhaps now my father, for the first time, was also asking what engineering is for and what his engineer son was going to do for Colombia upon return. Our conversations often included the question "What are you going to do when you come back?" and dreaming up plans of how and where I might secure a job in my native country. Not every answer or plan was acceptable. In my maternal family, there were three engineers who began their careers in import-substitution companies in the 1970s as salaried technocrats involved in industrial maintenance, agricultural field work, and metal work. They were incarnations of engineering for development yet their pathways were not an option for me. The status and content of their work were no match to the elite positions held by the Lucenas nor the kind of engineering that, according to ongoing geopolitical shifts, would help Colombia in the late 1980s.

With the Cold War coming to an end and a renewal of free-market ideology rising in the world stage under the name of *neoliberalism*, both rapidly shaping Colombian policies and politics, my father likely envisioned a kind of engineering that, enacted by his son, would bring technological development and innovation to Colombia to help it compete in world markets. So what could I do that was both acceptable to family and appropriate to country? How could I use my engineering degrees in a different way than that of a salaried technocrat? Could I bring my aeronautical engineering degree to Colombia to start new aircraft industries? As I continued to learn in my *Politics of Design* class about the complex politics and history that shaped U.S. aerospace, I realized that Colombia would never have a developed aerospace industry. Minimally, the development of aerospace required a historical moment (e.g., a transition from WW II to the Cold War), an intimate relationship among academia, government and industry (e.g., manifested in the U.S. military-industrial complex), huge public expenditures (only possible in rich countries like the U.S.), strong public support for them (a U.S. public fearful of communism), and large numbers of scientists and engineers to sustain it. Colombia had none of these. With no favorable external conditions for the development

---

[2] From the 1960s to the 1980s, my father occupied key government positions as private secretary to the President of Colombia, mayor of Bogota, president of Bogota's city council, and congressman. In all these positions, the ideology of development, as proposed by the U.S. for Latin America in the Alliance for Progress for example, was very pervasive in the formulation of domestic policies, programs, and government agencies in which my father was deeply involved.

of an aircraft industry and family elitism towards the professions weighing on my shoulders, I felt stuck.

After graduation, my search became more constrained when my father would only support graduate studies in Business Administration (BA) or Industrial Engineering (IE) because, according to him, these graduate programs combined with engineering would ensure higher status and salaries. Realizing that the Cold War was ending, that Colombia would not have an aerospace industry, and that free-market ideology was setting the rules for technological development, I reluctantly enrolled in BA and IE courses, including Operations Research(OR), the entry-level course for a master's in IE. Wanting to show my father that engineering could be applied to free-market development, I did my OR final project on personal-pager supply and distribution models for his new telecommunications business in Colombia. I developed a model to efficiently move pager inventory from Motorola USA to his business in Colombia. Throughout this course, humans are visible only to the extent that their inefficiencies need to be removed from system analysis.

At the same time, I began taking a course with Professor John Schumacher titled Philosophy and Engineering that deeply influenced my career trajectory. In this course, Robert Linhart's book *The Assembly Line* opened my eyes to the political and human dimensions of assembly line design and operations.[3] Becoming increasingly aware of managerial control of workers, I also wondered about the motives of engineers, "[W]as the intention of the engineers behind the design of the assembly line to construct an efficient system that would require such a hierarchical organization with a highly ruling authority?"[4] I could no longer continue in IE where naïve assumptions about humans in assembly lines reduce them to points along a production path and make them invisible. Again, I felt stuck since my father – my source of financing at the time—would only support graduate studies in fields that I could no longer undertake. Could engineering be applied to anything else other than the Cold War or the development of free-markets? I could not answer this question within the course offerings available to me in engineering. My engineering professors did not want to explore these questions with me. Office-hour sessions with them would quickly shift away from these larger issues to explanations about the variables in an equation. I had to change trajectory.

TO DEVELOP APPROPRIATE TECHNOLOGY FOR THE WORLD'S POOR?

Fortunately, a welcoming and nurturing group of faculty at Rensselaer's Department of Science and Technology Studies (STS) offered me a graduate assistantship to do ethnographic research on new engineering research centers. They were interested in my struggles and my insider knowledge of engineering labs on campus. This financing allowed me to pursue a master's degree in STS where I took courses such as the Sociology of Science, Values and Ethics, and Technological Development in Latin America. Influenced by these courses and their faculty, I was determined to find alternative answers to my question. I researched and wrote a course paper on appropriate technology (AT) as a

[3]Linhart, The Assembly Line, 1981.
[4]Linhart, *The Assembly Line*, 1981.

solution to poverty and hunger in the Third World. Assertively, I wrote in the paper's first sentence, "The purpose of this paper is to show that there is a kind of appropriate technology that, in fact, can alleviate the food crisis in Latin America."[5] Influenced by British economist E.F Schumacher's concepts of Intermediate Technology (IT) and his notion of *economics as if people mattered,*[6] I wanted to find out if there was a kind of *engineering as if people mattered.* In my engineering courses, I had become painfully aware that people did not matter. Although no longer the main source of financial support, my father was not out of the picture. He was also reading E.F. Schumacher's books and questioning the appropriateness of his ideas to Latin America. So he was challenging me to read the works of Chilean economist Manfred Max Neff whose work on "barefoot economics" was grounded in his direct experiences with poor communities in South America.[7] Perhaps resigned to the idea that I would not be an engineer developing free markets in Colombia, my father tried to ensure that my concept of appropriate engineering was at least appropriate to South America. During a summer visit to Colombia, my father arranged a meeting with Paolo Lugari, founder and then director of Gaviotas, an experimental farm for appropriate technology in Colombia,[8] with hopes I would secure employment with Mr. Lugari and thus finally return to Colombia. This did not happen as I was becoming increasingly skeptical of AT.

At this point in my search, I found that the normative calls for AT from philosophers, grass-roots social scientists, mainly economists, but none from engineers. I also found many encouraging examples of AT applied to food production in poor communities but could not find the engineering or engineers who designed and implemented these. Engineers were invisible to me in this history of AT.[9] Yet I was one of them and wanted to believe so hard that technology would solve the world's problems. Still early in my STS graduate program, I did not know how to understand the relationship between technology and society other than through *technological determinism*, a deeply-held idea by engineers that assumes that technology is the main cause of social change. In short, engineers build technologies according to internal logic of engineering and eventually these technologies bring social change. I did not know any better yet.

In a second research paper, I took a more empirical approach to finding the connection between engineering and appropriate technology. Langdon Winner's chapter "Building a Better Mouse Trap," where he argues that the AT movement became a fad mainly because it ignored the political realities of the time,[10] challenged me to test his claim empirically. I undertook fieldwork research in the US Northeast, visiting a number of AT experimental farms, AT-related NGOs, and government agencies. Unfortunately, Winner was right! The AT movement was politically dead, reduced to a number of showcase experimental farms and mail-order catalogs. The few engineers working for

[5] Lucena, Appropriate Technology, 1989.
[6] Schumacher, Small Is Beautiful, 1973.
[7] Max-Neff, From the Outside Looking In, 1981.
[8] See Weissman, *Gaviotas,* 1998.
[9] In 2008, I found out that the engineers who founded Volunteers in Technical Assistance (VITA) and tried to disseminate AT worldwide where located in Schenectedy, NY, only a few miles away from Rensselaer. Ironically, I never found them during this research and came to learn about them recently. See Williamson, "Small Scale Technologies for the Developing World," 2007.
[10] Reagan's conservative revolution and spread of free-market and neo-conservative ideas to many parts of the world. For a critique of AT, see Winner, *The Whale and the Reactor,* 1986, Ch. 6.

Appropriate Technology International (ATI was a part of USAID) would soon have to find work elsewhere as the Reagan administration eliminated federal support for AT. These research findings challenged my naïve views of engineers and technology. First, engineers working in non-profit areas such as AT are still heavily dependent on government and private funding. Second, engineering problems and solutions are never purely technical. I was learning that engineering problems and solutions always have political and economic dimensions that shape the way problems are defined and solved.[11] But I did not fully understand how. The relationship between engineers and government funding, and their location in relationship to the government, was becoming clearer yet discouraging. Third, without changing the underlying economic and political conditions, technology alone, no matter how appropriate, will not improve people's living conditions. If U.S. engineering had been primarily for the Cold War and for private industry, how could engineering and its technologies be for something else? And if possible solutions to improving people's living conditions rest on changing economic and political relations, how could engineers do that when they are not trained in economics or politics? I was stuck as my two fundamental assumptions about engineering and AT—engineers alone can develop AT; AT can solve the world's problems—and their logical conclusion–engineers can solve the world's problems—had been shattered.

ENGINEERING FOR ECONOMIC COMPETITIVENESS?

As part of my STS master's program, a graduate internship at NSF gave me a new opportunity to witness different dimensions of engineering. First, I worked for Rachelle Hollander in the Ethics and Values in Science and Technology (EVIST) program where I did a comprehensive study of NSF-funding of STS-related research including the field of engineering ethics. I learned that engineering societies and ethicists were trying to figure out how engineers should behave under special circumstances (e.g., conflict of interest, discovery of design flaws, etc.) yet they were not asking what is engineering for?

Next, I worked at NSF's Division for Engineering Infrastructure and Development (EID) alongside government officials, many of them engineers, who were constructing a new rationale to increase the number of engineering graduates under a post-Cold War rhetoric of economic competitiveness. At EID, I researched and wrote many documents and presentations on the engineering pipeline, a model and metaphor conceived by engineers at NSF to depict the problem of student recruitment, retention, and attrition in science and engineering.[12] As I helped program managers, division directors and the Assistant Director for Engineering collect data and create charts on demographic groups in engineering, I was beginning to realize that they were formulating a new answer to my pressing question. These engineers, now turned policy activists inside of NSF, rallied for increasing the number of women and minorities in engineering and funding for engineering R&D in new areas such as manufacturing, robotics, materials, and electronics in order to help the

---

[11] See Lucena, "Appropriate Technology," 1989.

[12] See, for example, National Science Foundation, *The Science and Engineering Pipeline*, 1987; Bowen, "The Engineering Student Pipeline," 1988.

U.S. compete economically in a post-Cold War world.[13] Although I did not know it at the time, these activists were developing a very compelling and successful argument that engineering was a key dimension of U.S. economic competitiveness and that recruitment and retention of women and minorities were key elements in this endeavor.[14] At EID, I connected with a network of engineering education activists who introduced me to my first conferences on engineering education and helped me research valuable data for my dissertation. Later, the significance of their activism would become visible in my book and the subsequent mappings that I did of engineering education. This network opened for me a new landscape of research and activism: engineering education.[15]

Wanting to theorize and conceptualize the shift from Cold War to economic competitiveness and its implications for engineering and engineers, I began looking for doctoral programs with faculty who were interested in this problem. Fortunately, I found the mentoring and guidance of Gary Downey at the STS program at Virginia Tech. There I experienced some degree of frustration as many STS faculty and students were more interested in science than in engineering. My main question— what is engineering for?—and its many variations often provoked yawns among most faculty who were exclusively interested in *science*. But I was determined to make engineering the focal point of my STS studies. For example, in my History of Science course, which emphasized the "scientific revolutions" in early modern Europe, I began looking for engineers. I realized that engineers were not visible due to a number of historical contingencies, mainly because engineering as a profession did not exist in most countries until the rise of the modern nation-state in late 18th century. Finding engineers here and there (e.g., military engineers in pre-Revolution France and craftsmen labeled engineers in pre-Industrial Revolution Britain), I had not developed conceptual and methodological frameworks to fully analyze them as historical actors. For my final paper, I chose to research and write about military engineering texts used during the American Revolution to guide the construction of fortifications and deployment of revolutionary troops. Although this research topic was a huge departure from the topics expected in this class, and even provoked disdain from my professor, I learned two key lessons about what engineering is for. First, French military engineers have developed theoretical approaches for conducting war and codified them in texts used in the French *Grandes Écoles* and later in other parts of the world. A clear and direct connection among the development of military engineering, engineering education and the strengthening of the French imperial state became visible to me. But then I did not have the theoretical sophistication to understand how this connection evolved and how much it came to characterize French engineering thereafter. Second, key American revolutionary leaders became determined to establish a connection between military engineering and the strengthening of the newly formed American state through the creation of West Point Military Academy and the U.S. Army Corp of Engineers. French engineering was heavily influential in this development and would come to play a significant role in U.S. engineering

---

[13] See, for example, National Science Board, The Role of the National Science Foundation in Economic Competitiveness, 1988.

[14] See Lucena, "Making Women and Minorities in Science and Engineering," 2000.

[15] In my dissertation, I researched and analyzed the history and politics of how different groups of engineers successfully developed the argument of "engineering for economic competitiveness" into programs and budgets for engineering education. See Lucena, *Defending the Nation*, 2005.

education. The French-sounding names of the equations in my engineering textbooks were beginning to make sense. Yet at the time, I did not know how influential French engineering has been around the world and how the French have a clear vision of what is engineering for.

I continued my journey through STS trying to apply theories, concepts, and methods from the social sciences to study engineering and engineers. Yet I was experiencing what Gary Downey later coined as the *downstream model*, i.e., the dominant assumption that science discovers truths and this knowledge is then applied in the form of applied science or technology and then put to use in society, often by engineers.[16] Most of my STS faculty and peers operated under the downstream model, often ignoring the contributions of engineers and finding engineering analytically uninteresting. I felt out of place during my first STS conferences where sessions on engineering were a rare exception and attracted only a few. All the excitement was in sessions led by big names in *science* studies. Not many in STS cared about engineering.

I found refuge in my participation in Gary's ethnographic research project on how engineering problem solving (EPS) challenges students as people.[17] This research would have a significant influence on my intellectual development vis-à-vis engineering and politics. For the first time, after six years of undergraduate engineering education and hundreds of homework sets using EPS, I began to understand EPS, the core method within the engineering curriculum, in cultural and political terms. Reading the works of Michel Foucault in Gary's course The Normative Structur[ing] of Science deeply influenced my thinking about how disciplining creates subjects. I began to understand how the disciplinary practices of engineering education (e.g., developing good studying habits, learning to follow the format of EPS, programming in FORTRAN) contributed to the formation of engineers as subjects. I was finding a different kind of answer to my question. In this case, engineering education is also about the formation of disciplined subjects (engineers) through the mastering of EPS. Those who resisted these disciplinary practices often had to leave engineering. But how about the disciplined subjects? What are they for? These revelations were both enjoyable and painful. After many years of being disciplined as an engineering student, I was now watching others struggling and resisting the disciplining of EPS, FORTRAN, and other practices in engineering education. As I followed a group of senior engineering students through their design project, they became another window into engineering education as they showed me how students articulate the tension between the constraints of EPS and the freedoms of engineering design. Thanks to my former NSF colleagues and these students, I could now see two sides of engineering that as a student were invisible to me: the arena of policy and reform at engineering conferences and the struggles of students with engineering curricula.

As I conducted both historical and ethnographic research on engineering education, a number of key questions emerged. First, is EPS universal? Is EPS used and considered as the core of engineering curricula everywhere in the world? If so, are similar disciplined subjects being formed [educated] in engineering schools all around the world? If not, then how are they being formed [ed-

---

[16] See Downey, "What is Engineering Studies For?" 2009.
[17] See Downey and Lucena, "Engineering Selves," 1997; Downey and Lucena, "When Students Resist," 2003.

ucated] in other parts of the world? Do engineers in different countries have the same relationship with the state as French engineers do? Conceptually and methodologically, how do we go about answering these questions? These were the key problems and questions that challenged me when Gary invited me to co-develop and co-teach the first version of Engineering Cultures at Virginia Tech in 1995.

FOR COUNTRIES AND NATION-STATES?

By 1995, calls for engineers who could be flexible and competitive in a global marketplace were filling the headlines of academic and professional magazines and newspapers. At the same time, cross-cultural competence emerged as a key desired characteristic for engineers who would help the U.S. regain and maintain its competitiveness in a post-Cold-War world.[18] Yet my participation in the development and delivery of Engineering Cultures was not motivated by a desire to help U.S. competitiveness but to engage students critically in understanding the historical and cultural roots and connections with engineering and, perhaps more importantly, to reflect on what these meant to their lives and career aspirations. The first offering to twenty-three students was marked by a heavy emphasis on engineering in France, Britain, and the U.S., as these three countries enjoy a significant amount of literature in English and strong historical connections in their developments of engineering. The course also included Japan as the country that the U.S. rhetoric and policies of economic competitiveness were responding to. Around that time Gary and I also completed a review of engineering studies that helped me become aware of the diversity of disciplinary perspectives and accounts on engineers and engineering.[19] Although Gary was the faculty of record for the course and ultimately responsible for its delivery and grading, a number of significant challenges emerged for me as a teacher in training.[20] For example, having just finished coursework in a STS program that was organized along disciplinary lines –history, philosophy, social sciences– I wondered about what kind of disciplinary perspective to privilege in this course. Historical? Anthropological? Sociological? Philosophical? Could one of these disciplines provide appropriate answers to the question, what is engineering for?

Pedagogically and conceptually, we wrestled with how to organize the content. Where do we begin the course? How do we organize a history of engineers? Do we begin in the past? Or do we begin at the present in a way that might be more appealing to students? If so, in what location at the present—Corporate? Academic? Military—and how far back do we dig into the histories of engineers? What other countries, besides Japan and those in Euro-America, should we include? How can we give visibility to areas of the world that have been ignored by mainstream historical accounts of engineering? How could we avoid making stereotypical generalizations of engineers and countries?

[18] For a comprehensive review and analysis of calls for flexible engineers, see Lucena, "Flexible Engineers," 2003.
[19] Downey and Lucena, "Engineering Studies," 1995.
[20] At this point, I begin to use "we" to refer to Gary and I as co-teachers of Engineering Cultures. However, the challenges and issues written about here are from my perspective only. I take full responsibility for the description and analysis of the course presented in this chapter.

We resolved some of these challenges during the course's first offering. First, we decided to begin with an account of the present by inviting students to see competitiveness as a cultural and historical problem, not as a given. This decision, of course, brought new challenges. For example, how could we provide students with a balanced perspective of what it is like to work as an engineer in the U.S. in the 1980s-90s without trivializing or demonizing organizational life and work? We chose Gideon Kunda's book *Engineering Culture* as an anthropological account of engineering work in high-tech organizations.[21] Personally, as an engineering graduate who never worked in an engineering organization, I found this account revealing and was glad that I did not have to choose a career pathway in the private sector. Yet students wrestled with Kunda differently. Some were excited about becoming engineers, earning high salaries and inventing cool gadgets. Others struggled with Kunda's depictions of managerial control of engineering work. Most were torn between these two extremes. I used the book many times in subsequent years and often wondered about its impact on students' careers five to ten years after graduation. Would they become aware of the power dimensions involved in engineering work? If so, would they be motivated to try to change these and perhaps begin asking what engineering is for? I still do not know.

Other challenges emerged during the unit on Japan. For example, how could we teach about Japan mainly from the perspective of American authors, many of them deeply influenced by the ideology of economic competitiveness and seeing Japan as a potential enemy? Neither Gary nor I are experts in Japanese history or culture yet we wanted to present to students a more balanced view of the complexities of Japanese society and avoid generalizations. How could we do this? Similarly, when developing and teaching the unit on the Soviet Union, I wondered how to present the history of engineering and the perspective of engineers in the USSR without demonizing an entire country or an entire ideology as has often been done in U.S. mainstream historical accounts of the USSR and socialism. We choose Loren Graham's book *The Ghost of the Executed Engineer*, a compelling account of an engineer's trials and tribulations as he was challenged by political turmoil and socialist and anarchist ideas during the governments of the Tsar, Lenin, and Stalin.[22] At the end, the account privileges the engineer as an individual hero while vilifying socialism as a political system. The engineer is executed by Stalin's regime.

I struggled with this representation since through my family I learned to see socialism under a more positive light and very different from Stalinism. Stalin was evil; socialism was not. The decade before I was born, my father had been a communist youth leader, an onsite witness of the Cuban revolution, and a student of socialist economics and politics. Throughout my childhood, I enjoyed many murals, paintings, and book covers beautifully crafted by my aunt, the official painter of the communist party.[23] I learned to sing "L'Internationale" before I knew that "Row, Row, Row Your Boat" even existed and, with the rest of the world, celebrated Labor Day on May 1st not in September. For my tenth birthday, I received the three volumes of Marx's *Das Kapital*. Thirty years later, I was

---

[21] Kunda, Engineering Culture, 1992.
[22] Graham, The Ghost of the Executed Engineer, 1993.
[23] See a small sample of these paintings at http://www.moirfranciscomosquera.org/public/Arte%20y%20Pueblo.htm. Also see my aunt's own writings on art and revolution in Lucena, *La revolución, el arte, la mujer*, 1984.

reading and teaching a book about engineers' grim fates under Soviet socialism. I was surprised to find out how little students knew about socialism and also how little I knew about personal tragedies under Stalinism. Teaching engineers was becoming a questioning of my own past.

When deciding on the content related to U.S. engineering in 20th and 19th centuries, we struggled with what historical events to include and with finding effective case studies to show students the historical development of engineering practice and education in the U.S. There are so many good historical case studies to choose from, each emphasizing different issues that we wanted students to know.[24] Yet we had to make difficult decisions. Should we include a case study showing the tensions and conflicts between design and manufacturing engineers? Or one showing attempts by engineers to make engineering more scientific when the profession was losing status? Or one showing how craftsmanship was replaced by mass production and the shift of engineering practice from the workshop to the corporation? Inevitably, every inclusion leaves other accounts out, an issue that continues to be problematic because students might never have another chance to learn these histories in their curriculum.

The histories of engineering in European countries were new to me. Learning and teaching about the histories of engineers in Europe challenged me on many fronts. First, I became painfully aware of my own ignorance of European history. As a young student in Colombia, a former colony of Spain, I learned the "official" history of Europe, a sanitized and linear account of historical events connecting Europe with the Americas in unproblematic ways. (Re)learning this history through the experiences of engineers opened my eyes to new questions. When and how does a country become a country? An empire? A colony? A nation-state? What role does the government, including its engineers, play in these developments? At the same time I felt oppressed by the dominance of Euro-American accounts. What was the development of engineering like in other parts of the world? Latin America? Africa? Asia? Why weren't students learning these? How does this imbalance contribute to Euro-American centric views of what counts as engineering, progress, development, and what does not? In short, I was struggling with my own lack of knowledge about these histories, my own immaturities as a teacher, my fears and anxieties as a foreign student, and my quest to find alternative answers to the question, what is engineering for?

At that time, my dissertation research challenged me for the first time to theorize the concepts of nation and state. These were unusual concepts in STS, usually perceived to be more appropriate for political science. Yet as I read accounts from engineers calling for a particular kind of curriculum or a new set of competencies to address changing needs of the U.S., I realized that the idea of nation invoked in the 1950s was very different from the one of the 1980s. So I began to wonder, what is a nation? Why have engineers been so preoccupied with responding to the needs and challenges of the nation? These questions would stick until much later when Gary and I began writing about engineers *and* countries, instead of just teaching about engineers *in* countries. I was finding yet

---

[24]See, for example, Reynolds, *The Engineer in America*, 1991.

another answer to my question. Engineering is for the nation. As images of the nation change so does engineering.[25]

My dissertation research also challenged me to research, interview, and listen to many engineers who have been actively engaged in invoking the nation in order to promote engineering in the 1980s and 1990s. Most of them were politically conservative, pro-American nationalists with a very different conception of the world than mine. Yet they were becoming acquaintances, sometimes friends, willing to trust me with their stories and giving me access to their social networks, files, and even homes. My quest to find what engineering is for was also challenging me to reconsider my view of people in relationship to their politics. Raised among socialist and liberal thinkers, I was now mingling with conservative engineers.

FOR CORPORATE COMPETITIVENESS?

ENGINEERING CULTURES AT ERAU

After taking my first faculty position at Embry-Riddle Aeronautical University (ERAU) in the department of humanities and social sciences (HU/SS), I quickly realized that the examples, case-studies, and even exam questions used in Engineering Cultures needed to be updated to the interests of my new audience: aerospace engineering students. I began learning how aerospace was mapped into the course's countries and how these countries were mapped into aerospace. During my first solo teaching experience, the course was fully loaded with content for I was anxious to show the course's value to my new colleagues and students. At the expense of student learning, I covered France, Britain, and Germany in only 3 weeks! Korea and China came next in just two more classes! I became guilty of trivializing the complex histories of countries for the sake of including countries that were becoming key sites in aerospace and relevant to U.S. economic competitiveness. (Boeing, for example, was procuring aircraft parts and systems from and selling aircraft to these countries.) I allowed the course content to be dictated by the supply chain instead of a commitment to proper coverage of the histories of engineers in these countries.

At the same time, this course was the source of tensions within my new department. Some of my HU/SS colleagues viewed me as a "sell-out" to corporate interests. In their eyes, the course's easy justification in an engineering school, particularly its close alignment with ABET accreditation criteria, made it suspect of co-optation by corporate interests. Others felt threatened by the course as it became an exemplar of how to integrate HU/SS with engineering. Often, HU/SS faculty educated in strictly disciplinary fields find refuge in maintaining a clear boundary between technical and non-technical courses. They feel safe in a non-technical domain and threatened when asked to address the problems and challenges of engineers. The Engineering Cultures course is clear evidence that the latter can be done from an interdisciplinary perspective rooted in HU/SS. Teaching this course was bringing me closer to engineers and farther apart from departmental colleagues.

[25] See Lucena, Defending the Nation, 2005.

In an NSF-funded CAREER grant, I tried to articulate the tension between present phenomena in aerospace (e.g., supply chain, mobility of technical workforce and knowledge across national boundaries) and the histories of engineers at aerospace production sites. Not fully understanding key concepts such as *nation-state* and *country*, I took *globalization* as a new conceptual challenge. Perhaps, this was intellectual irresponsibility, but I was determined to study globalization through the present and past of engineers. Researching the literature for this grant proposal in 1997, I realized that the fields of political science and international studies were not interested in engineers or engineering. To my knowledge, no one in those fields was asking, what is engineering for? Meanwhile, in STS, with very few exceptions,[26] globalization was not yet viewed as a conceptual and analytical challenge. And, as we have seen, engineers were treated as uninteresting actors downstream in the 'science-technology-society' linear model. I felt the need to break new analytical ground even if being conceptually ill-equipped to do so. I also felt lonely in a HU/SS department at a technical university where many faculty saw themselves as providing service courses to engineers, not as theoreticians of the connections between globalization and engineering. Yet I was working in a school where you could see these connections every day. Alumni, donors, advisory committees, and visiting speakers were clear examples of engineers working within the multiple supply chains of people, capital, and technology that exemplify the global economy.

Necessarily, the course became rich with these examples. As my research funding was granted and I accumulated social capital among aerospace executives and engineers, the doors of these supply chains opened wider to my inquiry. I discovered, researched and incorporated in the course examples about French and Mexican engineers developing avionics for Honeywell, British and Japanese engineers working at Boeing, and German engineers designing jets at Fairchild Dornier, to name a few. These examples complicated previous generalizations. In class, I could no longer simply state that "French engineers work for the government" when there were French engineers working at Honeywell, a U.S. private company. The challenge for me and my students was now to understand the French engineer's behavior and present practice in a for-profit company in the U.S. in light of his previous education in one of France's *Grandes Écoles*.

My fears of becoming co-opted escalated. I did not want to become a pawn of corporate interests at the expense of my students' education as critical thinkers. Yet the course's liveliness, richness, and relevance depended greatly on the mutual trust with executives in large private companies. I worked closely with a president of one of these companies, a very conservative pro-business Republican, who gave me almost unlimited access to meetings, people, and workshops, for he viewed my work as enhancing his company's competitiveness. I worked closely and actually became good friends with another senior executive who gave me access to his personal story, manufacturing plants, and key suppliers because he viewed my work as providing engineers with the proper education to succeed in his company. I needed them and it seemed that they also needed me.[27]

---

[26] For example, Schott, The World Scientific Community, 1991; Schott, World Science, 1993.
[27] For an overview of this research, see Lucena, "Globalization and Organizational Change," 2006c.

Their support for my course and research activities translated into political capital for the creation of a new program at ERAU. Based mainly in the humanities and social sciences, the new degree in Science, Technology and Globalization (STG) was about to face significant opposition in a university where technical curricula reigned supreme. But as one of the key premises of Engineering Cultures (i.e., there are significant cultural differences among countries that shape engineering practices) served as a justification for the new degree program, skeptical technical faculty, who first opposed the new program, welcomed it as they wondered if cultural differences also shaped other aerospace-related areas, e.g., air traffic control and management, aviation communication, manufacturing, passenger and flight crew behavior, pilot training, etc. HU/SS faculty, who previously opposed my course, came on board to teach in the STG program as they were willing to update their courses for a strategic sequence that led to graduation from a major. Yet as Engineering Cultures provided political and symbolic capital to the new degree program, my fears of co-optation to corporate interests continued.

The STG program came alive in 1999 with three focal areas: Technology Policy and Management, Aviation Ecology, and Security and Intelligence. In large part, I received tenure for my role in the conceptualization, development, and management of this new program. But after 9/11, the Security and Intelligence area became the main thrust of the program. As the fears about the outside world became pervasive in the U.S., administrators and students enacted these fears in problematic ways. The blind fears that emerged in the U.S. about Islam, the Middle East, and the Arab world began to influence the program in unexpected ways. University administrators saw 9/11 as an opportunity to attract homeland security funding, including proposals to incorporate new U.S. Air Marshall training in the STG program, while some students saw it as an opportunity to enact their fervent patriotism in the classroom. I decided to leave the program and the University, giving up my tenure and six years of work.

ENGINEERING CULTURES AT CSM

In 2002, I became director of the McBride Honors Program in Public Affairs for Engineers at Colorado School of Mines (CSM). I idealized this job as one where I could fully articulate and institutionalize the connection between engineers and politics; a place where I could go beyond asking the descriptive –what is engineering for?– to ask the normative –what *should* engineering be for? The potential connection between engineering and public service was both intellectually interesting and politically appealing. Sadly, I learned that a small group of faculty was determined to control the program and was not interested in providing historical, cultural, or political insights to engineering students who might one day work in public service. For two years, I lived in a conflict between directing the McBride program, which was supposed to be an exemplar in engineering education yet was controlled by faculty unwilling to make it relevant to engineers in public service, and teaching Engineering Cultures, a course quickly becoming an exemplar of innovation in engineering education.[28]

---

[28] Engineering Cultures received recognition from the Carnegie Foundation for the Advancement of Teaching and was instrumental in Gary and I securing Boeing Senior Fellowships at CASEE and an invitation to deliver a Distinguished Lecture at ASEE 2006 in

In its new home, Engineering Cultures had to go through significant revisions. First, the collection of case studies developed and used at ERAU did not resonate with chemical, petroleum, civil, and mining engineers at CSM. Aerospace is a foreign world to them. Geographically, students focus on locations very close to CSM (Colorado, Wyoming, Texas, Oklahoma) or very distant (Alaska, North Sea, Middle East). The node sites of aerospace supply chains, such as France, Germany, Britain, and Japan, and the practices of engineers at these sites, do not mean much to them. Institutional trajectory, student interest and an ongoing partnership between CSM and the Petroleum Institute in the United Arab Emirates clearly indicated that I needed to develop content for new countries and case studies in new industries. These realizations led me to submit, with Gary, a grant to begin developing content on engineers and engineering education in the Middle East by inviting engineers from the region to share histories of engineering education in their countries at the 2004 ASEE conference in Salt Lake City.

This project brought us new challenges. First, as we tried to define a region that had become highly visible after 9/11, we experienced the politics of regional inclusion. If we included Israeli guests, key Muslim and Arab authors found it problematic to participate. Some of them, willing to sit and share with Israeli peers, would not be granted permission to attend by their governments. If we defined the region as "Arab World," potential authors from Iran and even from Egypt indicated that they would not attend. If we defined the region as "Muslim World," we would be forced to consider potential authors from Indonesia and Malaysia and risked upsetting potential authors from Turkey who were trying to help their country make a case for European membership by making visible its official secularism. At the end, we ended up with a conference session and a publication entitled, Engineers and Engineering Education in Bahrain, Egypt, and Turkey.[29] Association with *countries*, rather than regions, proved more comfortable to everyone. Second, we faced the challenge of writing with and editing the work of author/engineers who often wanted to glorify and sanitize their history and the role of engineers in the making of their countries. For example, some authors did not want to acknowledge the connection between some engineers and fundamentalist Muslim groups. This project showed me how highly political new content development could be.

CHALLENGES

After resigning from the McBride Program at the end of 2004 because of the tensions described above, I offered Engineering Cultures as an elective housed at CSM's Division of Liberal Arts and International Studies (LAIS). There the course found resistance from faculty who claimed absolute ownership over the concepts of culture, nation-state, politics, and globalization and the relationships among these. The course came under intense fire and my second chance for tenure was in jeopardy. The course challenges narrow disciplinarians in a number of ways. First, it places engineers at the center of the development of nation-states and makes them visible as nation builders working in

Chicago. As we adopted comprehensive evaluation, developed multiple delivery formats (online, CDs, classroom), and engaged in cross-institutional (Virginia Tech vs. CSM) and multiple format (classroom vs. online) comparisons and evaluations, the course's visibility escalated in engineering education. See Downey et al., "The Globally Competent Engineer," 2006.

[29] Lucena et al., "From Region to Countries," 2006b.

ministries, national industries, universities, etc., in ways that some historians and political scientists find odd and problematic. After all, engineers have not been visible in the disciplines of history and political science. Second, it challenges traditional disciplinarians by constantly using history, anthropology, political science, STS, and literature to explain engineers. Some students who were closely associated with these disciplinarians wanted to take Engineering Cultures but were told that it would not count towards their plan of studies. One student wanted to develop an applied version of Engineering Cultures by enrolling in the European Project Semester in Denmark to test the lessons learned in Engineering Cultures while working alongside engineers from other countries. She was denied permission to count either Engineering Cultures or its applied version towards her coursework. Third, through a systematic and comparative deployment of assessment tools to evaluate students' knowledge, skills, and predispositions, the course challenges others to think about what constitutes evidence of student learning and development. I found out how uncomfortable some of my HU/SS colleagues can be when challenged by having to provide evidence of student learning. Fourth, the course challenges the often comfortable yet naïve assumption that the world of engineering is clearly divided between technical and non-technical dimensions. Operating under this assumption, HU/SS faculty often ignore the technical dimensions of engineering as something that takes place inside of engineering courses and for which they have no responsibility. Yet Engineering Cultures constantly bridges the technical and non-technical dimensions of engineering, and explores the historical and political contexts in which this split takes place. As an untenured faculty member teaching a highly controversial course, I decided to give the course "low visibility" within CSM and "high visibility" elsewhere by joining Gary in presenting papers, workshops, and lectures in many domestic and international professional and educational venues. Engineering Cultures proved to be both highly problematic and rewarding for my tenure case.

In spite of its innovation in content, delivery formats and evaluation, the course has many limitations that I have not been able to deal with successfully. For example, the course focuses mainly on engineering education institutions at the expense of sites of engineering practice. Students become hungry for descriptions and analyses of engineering projects, engineers in industry, engineering job market, etc. in countries where they aspire to travel and perhaps work. Often, I do not have rich examples for what they want. I am always wary of pulling examples of international engineering work from magazines and newspapers for these often trivialize the complexity of cross-cultural technical work. Also, the course's evaluation is limited to a 25-question quiz, a pre/post essay, and a student survey on how prepared they think they are for international work. This evaluation does not measure critical thinking or global competency on the job after graduation or co-optation to corporate goals. What if, after all, I am just preparing students to help companies improve their bottom lines and nothing more?

FOR HUMANITARIAN CRISES AND COMMUNITY DEVELOPMENT?

More recently, motivated by an emerging interest in humanitarian and community-development engineering in the so-called "developing world," I created a version of Engineering Cultures in the Developing World that includes Brazil, Colombia, and Mexico. This new version of Engineering Cultures challenged my students and me in new ways. For example, my research on Mexico challenged me to rethink the emergence of this country as a long-standing process of nation building followed by events of political independence and reaffirmation. Mexico's pathway to become a country is very different from Colombia's and Brazil's, a difference that can be seen in the development of their engineers and engineering education. In Mexico, I found an amazing diversity of engineering education institutions (e.g., Monterrey Tech, UNAM, IPN, Sistema de Tecnologicos), all positioned differently in Mexican history and politics. After taking this course, my students can no longer assume that all Mexican engineers are educated equally or view collaborations with U.S. engineers in the same way or all have the same desires and aspirations. They have learned that it is possible for engineers elsewhere to be motivated by diverse desires other than money and career advancement. Perhaps more important, my students no longer take for granted the stereotype of Mexicans as illegal immigrants trying to jump the fence to work in the U.S. but come to see Mexicans as people with a complex history, respectable engineering institutions, capable of forging their own history and images of progress, and with a unique answer to the question of what engineering is for. In Mexico, engineering is to build and serve the Motherland.[30]

Research in Brazil challenged me to learn the history of the only country in Latin America that became an Empire after a bloodless independence from Portugal. During my trips to Brazil, I felt ambivalence about interviewing engineers who played significant roles in one of the cruelest military dictatorships in Latin American history. Yet if I wanted to understand developments in engineering education and research in the last four decades, I had no choice but to earn their trust to interview them. At the same time, I made an unexpected and fascinating discovery of complex and diverse migration to Brazil, particularly Japanese immigrants who have become quite influential in engineering. Again, my students can no longer see Brazil through the stereotyping lens of "samba and soccer" but as a country with unparalleled regional and ethnic diversity in the Americas, infrastructural and industrial development of monumental proportions, and with its own conception of what engineering is for: order and progress.[31]

Even after being born and raised in Colombia, I found out that I knew little about the history of my home country. Reading and researching about Colombian engineers has challenged me to re-discover my Colombian self. For example, researching the development of engineering in Medellin, the industrial capital of Colombia, helped me understand the emergence of a protestant ethic in the region where my grandmother and father were born. I have begun to understand how protestant values permeated my traditional Catholic family. Reading the history of 20th century

---

[30] See Lucena, "De Criollos a Mexicanos," 2007.
[31] See Lucena, "Chasing Progress in Brazil," 2006a.

industrial development helped me understand my father's political trajectory, beginning with my grandfather's political career and their struggles during Colombia's most violent decade (1950s). I now understand their attitudes towards and struggles with my decision to study engineering in proper historical context. My students have also developed a new understanding of Colombia, not as a country of drug dealers and exotic plants, but as a country with a troubled political history that continues to shape its engineering and engineers to this day.

This new research on three Latin American countries coincided with the process of my application for U.S. citizenship. Interestingly, my own identity has undergone significant transformation. At a time that I was legally becoming more American, existentially I was becoming more Brazilian, Colombian, and Mexican as I researched, wrote and taught about the histories of engineers in these countries. For the first time in my life, I do not feel the emotional need to be physically present in Colombia as I find home in reading and teaching the histories of Brazil, Colombia, and Mexico. These transformations in identity and legal status led me to activism within engineering education in Latin America as I came to understand and value the perspectives and diversity of engineering colleagues in Latin America and the challenges they face by initiatives such as Engineering for the Americas to educate ABET-like engineers in Latin American engineering schools.[32] In most engineering education conferences, I experience ambiguity both as a participant observer and as an activist who wants understanding and respect for the marked regional, ethnic, and institutional differences among and within countries and their engineers. My network of engineering friends in Latin America includes those who oppose the Engineering of the Americas and those who support it. Over the past five years, I have found myself traveling with, consulting for, and visiting deans of influential schools of engineering in Brazil, Colombia and Mexico, helping them figure out how to align themselves with opportunities for engineering education reform coming out of the U.S. and Europe. At the same time, I also confer with U.S. engineering education reformers who are trying to figure out where and how in Latin America to implement new initiatives for reform.

Engineering Cultures in the Developing World is now part of a minor in Humanitarian Engineering at CSM. The three key questions at the heart of Engineering Cultures – What is engineering for? What constitutes engineering knowledge? What are the expectations on engineers' service to society?— have also served as inspiration for new courses on Humanitarian Engineering Ethics (HEE) and Engineering and Sustainable Community Development (ESCD).[33] As engineering faculty and students at CSM and elsewhere become interested in applying their knowledge and skills in humanitarian and community development situations, I have become increasingly preoccupied with their uncritical posture regarding whether engineering is an appropriate approach for such endeavors. Most engineering faculty and students involved in these matters make a number of problematic assumptions. For example, they assume that engineering as currently taught and learned in curricula rooted in the Cold War, and in some cases updated for economic competitiveness, can be easily applied to humanitarian and community development circumstances. Second,

[32] See, for example, Scavarda et al., "The Engineer of the Americas," 2005.
[33] See Lucena et al., "Theory and Practice of Humanitarian Ethics in Graduate Engineering Education," 2007.

they assume that an engineering education obtained in the U.S. coupled with a "desire to help" gives them the intellectual and moral authority as experts to solve the problems of those they consider to be in need. They never question whether this attitude could actually reinforce paternalism towards people in poorer countries. Third, they assume that design approaches for private industry, where there is a clear client-expert relationship defined mainly in legal and budgetary terms, can be easily adapted to new humanitarian and/or communitarian situations. Fourth, most assume that technologies designed and developed in a particular location can be easily transferred to another location. Motivated by these problematic assumptions, I am committed to challenge engineers to critically reflect on whether engineering problem solving (EPS) and engineering design methods are appropriate approaches to humanitarian and community development problems. I want them to question whether expectations of service to their societies as engineers coupled with desires to help, often motivated by religious missionary views, is all they need to try to solve other people's humanitarian and development problems.[34]

As new courses in humanitarian engineering adopt the central questions and method of Engineering Cultures—to map and listen to different perspectives—they also bring new challenges to Engineering Cultures. For example, the course's main method –Problem Definition and Solution (PDS)—calls for mapping people's perspectives, or their location, knowledge, and desires, as they interact in defining and solving problems. As my students have learned to apply this approach in a number of situations, they have also pointed to limitations. For example, students have discovered that a perspective could also include a material dimension beyond location, knowledge, and desires. As people come together around a problem, they also bring artifacts and systems already built. Students also have challenged the assumption that applying PDS necessarily results in non-conflictive mediation. Often, they have come to recognize that conflict is unavoidable.

New conceptual challenges also emerge as I research engineers' activities beyond the boundaries of a country. What does it mean when engineers' commitments extend beyond their country to include transnational challenges such as international development or humanitarian relief? I am beginning to understand and write about what engineering is for in relationship to development, sustainability, and community.[35]

A more significant challenge came from a community activist from Mexico who challenged us to question first whether "our struggles" are the same as his community's struggles before we can even sit and talk about how engineers can help his community. Are we as U.S. engineers interested in helping the people of Chiapas, Mexico, in their struggle to recover ancestral lands? Or are we just interested in problem solving in a community outside of the U.S. as a means to enhance students' international experience? He challenged us to engage and empower a community to decide what engineering is and could do for them. In short, this community activist helped me realize that in community development a community, not the engineers, should decide what engineering is for.[36]

[34] See Schneider et al., "Engineering to Help," 2009.
[35] See Lucena and Schneider, "Engineers, Development, and Engineering Education," 2008.
[36] See Schneider et al., "Where is 'community'?" 2008.

THE PROBLEM OF SCALE

Throughout this journey, key issues have prevented Engineering Cultures from disseminating beyond its current offerings at CSM and Virginia Tech. First, Engineering Cultures is an elective course that has to compete for student enrollment with other electives which might be more conveniently scheduled or enjoy a reputation of being easier among students. How can we make Engineering Cultures attractive in institutional cultures where many students still seek easy humanities courses with minimal requirements for reading and writing? Furthermore, many engineering students seek efficient ways to prepare themselves for international travel. Often they prefer brief introductory sessions focusing exclusively on the places where they are traveling to semester-long courses such as Engineering Cultures where the countries covered might not be the ones they will be visiting. Yet most students who take the course recognize that its methods and approach to learning about engineers are applicable to study and travel to other countries.

Second, after learning about the course through workshops and publications, interested faculty are very enthusiastic about the course content yet apprehensive about teaching the course in their own institutions. Gary and I have developed unique expertise in teaching this course and have the advantage of years of reading, research, and writing on Engineering Cultures. Gary has successfully trained STS graduate students who will probably teach it elsewhere after graduation. But I am not sure if a graduate student who is not trained by either one of us can successfully teach the course, or if a faculty member with a course syllabus, a reading list, and some training in social sciences can teach the course.

Third, we proposed an Engineering Cultures textbook to a number of publishers in engineering education who have been reluctant to consider the idea. The world of publishing in engineering education reflects the dominant categories of the engineering curriculum: engineering sciences, basic sciences, math, and design. There are a few titles on writing, communication, engineering economics, and engineering ethics. Elective courses such as Engineering Cultures might be risky to publishers. I am not sure yet how to deal with this challenge.

Fourth, I have experienced resistance from HU/SS faculty to what Engineering Cultures stands for. Some view Engineering Cultures just as a vehicle to co-opt students to corporate interests by teaching them about cultural differences and helping them become more effective employers who can increase profits around the world. They do not understand that ultimately Engineering Cultures is about helping students understand the strength and limitations of their own perspectives while valuing those of others. Others do not understand the course's attempt to blur the boundaries between the technical and non-technical dimensions of engineering. These faculty take comfort in leaving those boundaries untouched and invite engineering students to the non-technical side to "take a break" from engineering. As described above, other faculty critique Engineering Cultures' interdisciplinary approach and its alternative explanation of concepts that have been the purview of political science. So quite often, my participation and ownership of the course isolates me from mainstream conversations in my own department.

As of this writing, my quest for answers to the question, "What is engineering for?" continues to shape new scholarly interests on engineering, social justice, and immigrant engineers, and it influences my current role as co-Editor of the journal *Engineering Studies*.[37] Unsure where this quest will take me, I take comfort in knowing that there are many committed people trying to construct alternatives to engineering for military and free-market development. For now, my commitment rests on helping these activists map the complex landscapes of engineering education and practice at different times and places.

REFERENCES

Bowen, J. Ray. "The Engineering Student Pipeline: An Introduction." *Engineering Education* 78, no. 8 (1988): 733–734. 366

Downey, Gary Lee and Juan C. Lucena. "Engineering Studies." In *Handbook of Science, Technology, and Society*, edited by Shelia Jasanoff, Gerald E. Markle, James C. Petersen, and Trevor Pinch, 167–188. Thousand Oaks: SAGE, 1995. 369

Downey, Gary Lee. "What is Engineering Studies For? Dominant Practices and Scalable Scholarship." *Engineering Studies* 1, no. 1 (2009): 55–76. DOI: 10.1080/19378620902786499 368

Downey, Gary Lee and Juan C. Lucena. "Engineering Selves: Hiring into a Contested Field of Education." In *Cyborgs and Citadels: Anthropological Interventions in Emerging Sciences and Technologies*, edited by Gary Lee Downey and Joseph Dumit, 117–142. Sante Fe, NM: SAR Press, 1997. 368

Downey, Gary and Juan C. Lucena. "When Students Resist: Ethnography of a Senior Design Experience in Engineering Education." *International Journal of Engineering Education* 19, no. 1 (2003): 168–176. 368

Downey, Gary Lee, Juan C. Lucena, Barbara Moskal, Rosamond Parkhurst, Thomas Bigley, Chris Hays, Brent K. Jesiek, Liam Kelly, Jane Lehr, Jonson Miller, Sharon Ruff, Jane Lehr, and Amy Nichols-Belo. "The Globally Competent Engineer: Working Effectively with People Who Define Problems Differently." *Journal of Engineering Education* 95, no. 2 (2006): 107–122. 375

Graham, The Ghost of the Executed Engineer, 1993.

Kunda, Gideon. *Engineering Culture: Control and Commitment in a High-Tech Corporation*. Philadelphia: Temple University Press, 1992. 370

Linhart, Robert. *The Assembly Line*. Amherst: University of Massachusetts Press, 1981. 364

Lucena, J.. "Militarization of Technology," 1987.

[37]http://www.tandf.co.uk/journals/titles/19378629.asp

Lucena, Clemencia. *La revolución, el arte, la mujer*. Bogotá: Editorial Bandera Roja, 1984. 370

Lucena, Juan. *Appropriate Technology: A Contemporary Review*. Troy: Rensselaer Polytechnic Institute, 1989. DOI: 10.1177/0270467603259875 365, 366

Lucena, Juan. "Making Women and Minorities in Science and Engineering: Nation, NSF, and Policy for Statistical Categories." *Journal of Women and Minorities in Science and Engineering* 6, no. 1 (2000): 1–31. 367

Lucena, Juan. "Flexible Engineers: History, Challenges, and Opportunities for Engineering Education." *Bulletin of Science, Technology, and Society* 23, no. 6 (2003): 419–435. DOI: 10.1109/MTAS.2006.1649022 369

Lucena, Juan. "Chasing Progress in Brazil: Engineering Education for 'Ordem e Progresso'." Paper presented at the VI Jornadas Latinoamericanas de Estudios Sociales de la Ciencia y la Tecnologia, Bogota, Colombia, 2006. DOI: 10.1080/07341510701300361 377

Lucena, Juan, Gary Lee Downey, and Hussein A. Amery. "From Region to Countries: Engineers and Engineering Education in Bahrain, Egypt, and Turkey." *IEEE Technology and Society* 25, no. 2 (2006): 3–10. 375

Lucena, Juan. "De Criollos a Mexicanos: Engineers' Identity and the Construction of Mexico." *History and Technology* 23, no. 3 (2007): 275–288. DOI: 10.1080/03043790802088368 377, 378

Lucena, Juan, Carl Mitcham, Jon Leydens, Junko Munakata-Marr, Jay Straker, and Marcelo Simões. "Theory and Practice of Humanitarian Ethics in Graduate Engineering Education." Paper presented at the 114th ASEE Annual Conference and Exhibition, Honolulu, HI, United States, 2007.

Lucena, Juan and Jen Schneider. "Engineers, Development, and Engineering Education: From National to Sustainable Community Development." *European Journal of Engineering Education* 33, no. 3 (2008): 247–257. DOI: 10.1080/03043790600644040 379

Lucena, Juan C. *Defending the Nation: US Policy making in Science and Engineering Education from Sputnik to the War Against Terrorism*. Lanham, MD: University Press of America, 2005. 367, 372

Lucena, Juan C. "Globalization and Organizational Change: Engineers' Experiences and Their Implications for Engineering Education." *European Journal of Engineering Education* 31, no. 3 (2006): 321–338. 373

Max-Neff, Manfred. *From the Outside Looking In: Experiences in Barefoot Economics*. Uppsala, Sweden: Dag Hammarskjöld Foundation, 1981. 365

National Science Board. *The Role of the National Science Foundation in Economic Competitiveness*. Washington, DC: NSF, 1988. 367

National Science Foundation, Policy Research Analysis Division. *The Science and Engineering Pipeline.* Washington, DC: NSF, 1987. 366

Reynolds, Terry. *The Engineer in America: A Historical Anthology from Technology and Culture.* Chicago: University of Chicago Press, 1991. DOI: 10.1109/MTS.2009.935008 371

Scavarda do Carmo, Luiz C., Lueny Morell, and Russel C. Jones. "The Engineer of the Americas." Paper presented at the 2005 ASEE Annual Conference and Exhibition, 2005. DOI: 10.1080/03043790802088640 378

Schneider, Jen, Juan C. Lucena, and Jon A. Leydens. "Engineering to Help: The Value of Critique in Engineering Service." *IEEE Technology and Society Magazine* 28, no. 4 (2009): 42–48. 379

Schneider, Jen, Jon Leydens, and Juan Lucena. "Where is 'community'? Engineering Education and Sustainable Community Development." *European Journal of Engineering Education* 33, no. 3 (2008): 307–319. DOI: 10.1177/016224399301800205 379

Schott, Thomas. "The World Scientific Community: Globality and Globalization." *Minerva* 29, no. 4 (1991): 440–462. 373

Schott, Thomas. "World Science: Globalization of Institutions and Participation." *Science, Technology, and Human Values* 18, no. 2 (1993): 196–208. 373

Schumacher, E.F. *Small Is Beautiful: Economics As If People Mattered.* New York: Harper Torchbooks, 1973. 365

Weissman, Alan. *Gaviotas: A Village to Reinvent the World.* White River Jct.: Chelsea Green, 1998. 365

Williamson, Bess. "Small Scale Technologies for the Developing World: Volunteers for International Technical Assistance, 1959–1971. Paper presented at the Society for the History of Technology (SHOT) Annual Conference, Washington, DC, United States, 2007. 365

Winner, Langdon. *The Whale and the Reactor.* Chicago: University of Chicago Press, 1986. 365

CHAPTER 18

Location, Knowledge, and Desire: From Two Conservatisms to *Engineering Cultures* and Countries

Gary Lee Downey

"[A]s I began to mature, both emotionally and academically, I suddenly felt confined by the requirements of my [engineering] curriculum."
Author, Essay for scholarship competition, 1973.

I have just finished teaching a semester of *Engineering Cultures* with Grace Hood, Jongmin Lee, and Nicholas Sakellariou, graduate students in Virginia Tech's Science and Technology Studies (STS) Program. The main purpose of this large, 150-student course is to help engineering students become critically aware of their own knowledge and commitments and analyze them in relation to those of others, including both engineers and non-engineers.[1] Its main pedagogical device is a voyage across space and time. Class modules in this version examined the emergence of engineers across the territories of Britain, France, Germany, Japan, the Soviet Union, and the United States. Role-playing exercises challenged students to describe and perform perspectives other than their own. Analyses of recent reform movements in the United States sought to help students reflect on what led them into engineering, what is currently at stake in engineering curricula, and who has stakes in the contents of engineering knowledge.

Practices for helping students map their knowledge and commitments that I call "location, knowledge, and desire" sought to prepare them to grapple intelligently with conflict. These practices challenge students to ask of others: How are they located? What do they know? What do they want?

Asking these questions is designed to help students recognize and critically analyze how *they* are located, what *they* know, and what *they* want. They help students see that engineering practices always have normative contents, which is to say they are always linked to broader social projects. They help students recognize that other people with stakes in engineering work will likely

---

[1]The elective course regularly includes 5-10% non-engineers. We modify exercises to fit their pathways.

be located differently, have different configurations of knowledge and understandings of their work, and different desires for ultimate outcomes. Students used these questions to assess their pathways into engineering in the two-part Reflections Assignment. They are also questions I have long asked about myself.

The course has gained a reputation among engineering educators within my institution and beyond as helping students prepare for the so-called global workplace. These colleagues usually mean it can help engineering students get jobs in companies that have gone multinational. Indeed, it can. But I am greatly disappointed when its contributions to a student's learning appear to go no further.

I developed these questions and the contents to begin answering them as a first, crucial step in helping students overcome dominant practices in the core of engineering curricula that are, in my judgment, self-limiting. These practices tend to confine engineering service to the narrow channel of technical support—solving technical problems defined by others.[2]

Introducing students to others elsewhere in the world who define problems and understand their practices differently, including both engineers and non-engineers, is a strategy for challenging them to more fully understand and challenge themselves. When the process works well, it helps students analyze and assess the larger purposes and broader implications of their work as engineers in relation to the work and commitments of others.[3] The course is part of a larger pedagogical project designed to make critical analysis of one's own knowledge and commitments an integral part of engineering education and practice.[4]

I have come to view the larger project as strengthening engineering service. Self-limiting, conservative, narrow, or weak engineering service emphasizes the sharply restricted activity of solving problems for others through technological design without questioning or analyzing the contents of those problems or the effects of their solutions. Restricting engineering education to preparing students to solve problems defined wholly by others can limit their abilities to see, let alone engage, possibilities for stronger engineering service.[5] By confining engineering service to technical support, it also contributes to limiting the visibility and influence of technical practitioners who define their knowledge as in service to others.

Strong engineering service, by contrast, expects and invites discussion and debate over the extent to which producing and using engineering knowledge actually serves broader social projects and what those projects might include. Its practitioners understand it as having both technical and nontechnical dimensions, with broader commitments that are necessarily and explicitly multiple. Integrating discussion and debate about the contents of engineering service work can help engineers both recognize and assess those projects that are dominant or widely taken for granted and improve their abilities to develop and scale up alternative configurations. Yet strong engineering service is all too often a mere fantasy. Over the years, I have come to view the integration of critical self-analysis

---

[2] Downey, "Keynote Address," 2005.
[3] Downey, "The Engineering Cultures Syllabus as Formation Narrative," 2008; Downey et al., "The Globally Competent Engineer," 2006.
[4] Downey, "What Is Engineering Studies For?" 2009.
[5] Downey and Lucena, "Engineering Selves," 1997b.

into routine engineering practices as necessary to overcome the self-limitations in engineering work that contribute to the invisibility of engineers.

In this personal geography, I recount the travels that led me to develop practices for helping students understand and critically analyze their interests and identities as prospective engineers. I begin by describing my initial unquestioned acceptance of narrow engineering service and early rumblings of discontent. I then trace a flying leap into cultural anthropology and practices of external criticism, which ultimately came to feel as self-limiting and conservative as I had found my engineering curriculum to be. Wrestling with two issues fueled an interest in going beyond studying centers of power in science and technology to attempting to participate critically within them. One was a developing insistence that no research or teaching is ever innocent, in the sense that it purely describes or informs, because it always has meaning in relation to dominant practices beyond the academy. The other was a growing awareness that because I am endowed with many privileges I can never claim to be an innocent outsider, a pure critic. After a foray deep into federal policy making about radioactive waste management, pursuing these issues led me to a career in Science and Technology Studies (STS). There, my frustration with the relative invisibility of engineers and engineering led me to realize that overcoming self-limitations in engineering service had been my main interest and goal all along.

MY FIRST CONSERVATISM: ACCEPTING THE DOMINANT

In April 1970, I was a freshman at Lehigh University completing preparatory courses for a mechanical engineering curriculum that would begin the following year. A week of student protests had just led the administration to cancel classes for the following week.[6] I was mostly glad to have extra time to do my homework.

The protests at this all-male institution were against control by establishment authority that appeared to be both absolute and homogeneous. Students were left out, made officially impotent. "Does Lehigh Have You By the Balls?" read the signs calling us to the "Power Structure Forum." The daily events of protest, debate, and discussion began April 7, attracting up to 2,000 students to plenary events (total enrollment was 3,000). On April 15, the confrontations culminated in agreement over a new power-sharing arrangement. A "Lehigh University Forum" was formed with organizational authority below the Board of Trustees. Its 125 members included sixty each from the faculty and student bodies and five from the administration, including the President (notably giving faculty and administration a majority).[7] The Forum had authority to set policy on a range of issues involving "[a]cademic program and planning," "[a]cademic environment," and "[e]xtracurricular activities." It was entitled to review "[l]ong-range planning," "overall budget," "[c]ommunity relations," and "[a]ll administrative appointments at the rank of Dean and above." If it disagreed with the president on these matters, "both such positions shall be presented to the Board of Trustees." Finally, it could review and make recommendations on "[c]urriculum, [r]esearch policy, and [a]cademic discipline."

[6] Cole, "The Rights of Spring," 1970.
[7] Forum, "Constitution of the Lehigh University Forum," 1970a.

In Fall 1970, one of its first actions was to eliminate Saturday classes.[8] A plan to go co-ed in 1971 was evidently already near approval when the protests began. The faculty approved the new Forum constitution April 27 and the Board of Trustees May 2.[9] The Kent State shootings took place May 4. On May 8, 100,000 people arrived in Washington, D.C., 185 miles from Lehigh, to protest the shootings, the military incursion into Cambodia, and the Viet Nam war itself.

Although I attended several of the forums and meetings at Lehigh that week, I did not understand what was taking place. Or more importantly in retrospect, I could not feel what was taking place. I certainly enjoyed watching students courageously defending radical professors and confronting the university president and other administrative leaders. I did not, however, feel their anger. I greatly admired its clarity and certainty but did not share the profound sense of impotence or find hope and liberation in the student demand for participation in governance. It never occurred to me to go to Washington to participate in the anti-war protest. I could not afford such a trip, and final exams were coming.

I had effectively become an engineer long before finishing high school. When I was collecting letters of recommendation for college, I told my public high school physics teacher, Dr. Speer, I was uncertain about whether to become a mechanical engineer or chemical engineer. "Are you sure you want to become an engineer?," he asked. I was stunned. Of course I was going to become an engineer. What else was there?

I was literally unable to picture a future life course that did not pass through engineering. I had taken this for granted for many years. I was going to become a first-generation college student from Pittsburgh, having been raised in rented houses and supported by my father's work as a salesman and my mother's work at home. I was good in math and science. Only many years later did I reflect on the fact that my free and independent choices had both gender and racial dimensions. Of fourteen cousins, I was one of five boys who headed off to become engineers. All nine girls became teachers or nurses (well, one in hospital nutrition). Most of my uncles worked in blue-collar trades and all my aunts worked at home. One uncle had made it by going to college and becoming a civil engineer. This entire world was white. Its main internal divisions were between Protestants and Catholics and among descendants of German, Irish, and Eastern Europeans. I had no expectation of encountering black people in college (and I mostly did not).

I found myself in an expensive private institution because they gave me a full scholarship for tuition ($2,000/year). The other key reason was its location within a half-day's driving distance of Pittsburgh and not in either Ohio (an alien territory defined by allegiance to the Cleveland Browns[10]) or West Virginia (hillbilly jokes). To pay room and board, my mother went to work as a teacher's aide in an elementary school. The day my father dropped me off at Lehigh was the first time I saw the

---

[8] Forum, "Motions Passed by the Lehigh University Forum," 1970b.

[9] See also Yates, *Lehigh University*, 1992.

[10] For a brief period during the Depression, my maternal grandfather was part-owner of the Pittsburgh Steelers. The connection became genetic.

campus. I was there for an engineering degree, eventual white-collar job in industry, upward class mobility, and someday to assume responsibility for (over) a woman and children.[11]

In my 1970 world, control of curricular requirements by an undifferentiated engineering faculty, administration, and University, with non-engineering faculty in indirect support, was a natural state of affairs.[12] Freshman year was about completing preparatory courses, and by April I knew I would join the mechanical engineering department in Fall semester. I would learn then without surprise that the lead author of our textbook in Engineering Statics, Ferdinand Beer, was the department head. I did not know what we called "Beer and Johnson" was becoming, or already was, the leading statics text in the country. It was just our book and my job was to learn all the permutations of "sum of the forces equals zero."[13]

I greatly enjoyed learning the engineering sciences during my years at Lehigh. I believe the pleasure was in part about gaining confidence through control. One did not simply take classes in statics, dynamics, kinematics, thermodynamics, or mechanics of materials. Each was an experience that demanded initial submission and posed a challenge to gain control.[14] You could be confident of control if you earned A's. I earned A's. In each class, I entered a world that was both internally clear and externally powerful. Each time I faced anxiety about whether or not I could master it. Each time I did, it added to a sense of strength and promise for my future.

Not until years later would I realize just how congruent these cycles of submission and control were with the cycles of weakness and strength I experienced as a white boy and man—in which you are only as good as the last time you proved yourself. I simply accepted my job was to prove myself as a person, whether on a statics test or a basketball court. I did not realize it was a privilege to believe I could fit anywhere I chose to go as long as I proved myself as an individual. I was not inappropriate biologically in any public place.

Enjoying engineering was also about the certainty of serving. Engineering was a pathway for improving the world through technical work. The specifics were not analyzed or analyzable. I was not a tinkerer, so I was not imagining myself as a genius inventor. I accepted and embraced the achievement of gain through mathematical problem solving as the core vehicle for future contributions.[15]

I also took it for granted my pathway was leading to a career in private industry. This was wholly unproblematic. Good things came from industry, including good jobs and a good life. I gladly accepted an invitation from my uncle to work on construction projects in a steel mill during two summer breaks. I was delighted to win a summer position at Eastman Kodak in 1972 whose high pay and many learning opportunities in photography were geared toward attracting me back after graduation. To fulfill a research requirement in a course on air pollution engineering, it was a given

---

[11]This fits the typical profile of the engineering student described in Gerstl and Perrucci, *Profession without Community*, 1969.

[12]As a graduate student in cultural anthropology, David Schneider, the kinship theorist, told me, "Whenever you hear the word 'natural,' change it to 'cultural' and see how it changes everything." He also taught me to change "seminal" to "ovular."

[13]Beer and Johnston, *Mechanics for Engineers*, 1962.

[14]I have analyzed this iteration in Downey, *The Machine in Me*, 1998b, 134–209, especially 139–141.

[15]On the achievement of gain through mathematical problem solving in the engineering sciences, see Alder, *Engineering the Revolution*, 1997, 60.

I should seek the help of researchers in U.S. auto companies.[16] I was not surprised when my final exam in Ocean Engineering began, "Your firm has decided to enter the offshore business." It then asked us to write yet another consulting report, this time explaining what staff, reference materials, and equipment the firm should obtain as well as the sorts of contracts it should initially seek. We were reminded of the need to be cautious "until a reputation is established" and that of course "the company cannot long exist without making a profit."[17] For me, industry was just a place to do engineering work and I fully expected to move from company to company to build a career.

While proving myself in mechanical engineering, however, I also began to experience the curriculum as confining, as the epigraph above from 1973 indicates. To have a sense something is limited, one has to be able to see beyond its boundaries. The main vehicle for my resistance was a developing ability to see things from different points of view.

One night in eleventh grade, I had had a dream entirely in Spanish. I had been studying Spanish since the seventh grade and would be approaching working fluency when I finished high school. I experienced the dream as a wholly altered reality. I had joined a different world. People and things were different when I engaged them through the lens of Spanish. At the time, I was mostly proud I could have the dream at all. I was gaining mastery of the language. But it also triggered a developing interest in perspective and point of view. What most struck me is that the language itself seemed to have a point of view. I was seeing the world differently. This stuck with me. When I later sat watching parades of passionate speakers at the protest events in April 1970, I spent much of the time trying to figure out who they were, how they got to be where they were, and just how they viewed themselves in the world. I saw the forums as clashes of perspectives.

One reaction to my increasing recognition of different perspectives was a developing obsession with the question: What is my perspective? I tended to phrase it spatially: *Where* do I stand in all of this? I continued enrolling in Spanish courses, and later French courses, steps that wreaked havoc with my engineering class schedule. I found no other engineers within them. I was also jealous of all those students walking around campus who were learning philosophy and literature. I wanted those experiences as well. Reading had never been part of my daily life as a child. My main mentor in philosophy and literature was my girlfriend June, an English major at a nearby women's college. She introduced me to feminist literature (which I only browsed until Marilyn French's 1977 novel *The Women's Room*[18] grabbed me hard and forced me to begin interrogating rigorously the privileges I gained from being male). I began enrolling in classes in linguistics (to figure out just how languages work) and anthropology (to become informed about perspectives that lived in other parts of the

---

[16] Choosing to study the effects of combustion chamber design on auto emissions, my innovation was to cross the boundary separating mechanical from chemical engineering. My letters requesting support asked for information on "design criteria" and any "experimental data" they might be willing and able to share with me (they were not). See Downey, "Letter from Gary Downey to G.J. Huebner, Jr.," 1973c.

[17] Richards, "*Ocean Engineering* Final Examination," 1973.

[18] French, *The Women's Room*, 1977.

world).[19] Also, experiences with marijuana and, especially, mescaline became enormously attractive as pathways to altered realities, along with searching Carlos Castaneda books for indicators of deeper meaning (I found none).

What attracted my attention more than anything else, however, was a series of emerging controversies over energy and environmental issues, sharpened especially by the so-called Energy Crisis of 1973. To me, these were the 60's issues of the 70's. Perhaps unquestioned service through engineering work in the private sector was not unproblematic after all. Engineering was involved in producing technologies, and people were fighting over the impacts of technologies. What did that mean for engineers? I consumed popular articles and books about debates over such things as the supersonic transport, anti-ballistic missile, clean air and water, and, most prominently, nuclear power. As I read these literatures, I found I could understand the technical arguments advocates made on both sides of each debate. This was a good thing. Through engineering science courses, I was indeed gaining knowledge that could prove helpful. But the mathematical tools for problem solving that I was acquiring in mechanical engineering were clearly insufficient to understand controversies over energy and the environment.

To better connect my technical learning to my interest in these debates, I persuaded the associate department head to allow me to adapt my mechanical engineering degree to include environmental courses (the term "environmental engineering" did not yet exist in my world). I enrolled in the *Air Pollution and Ocean Engineering* class, as well as *Waste Water Control*, *Theory of Internal Combustion Engines*, and *Nuclear Reactor Engineering*, while bypassing a required course in *Heat Transfer*.

Tweaking my engineering pathway did not feel sufficient, however. While I understood the technical issues at stake in energy and environmental conflicts, what I could not understand was why people were fighting. What was at stake among them? I was still attending the Power Structure Forum in Grace Hall. The issues had just shifted. How could I go through life without understanding them? Might I be fooling myself about improving the world through engineering? Where did I stand in relation to these people?

One late night in the library I wrote myself a letter about this growing struggle. In the voice of a 1970s twenty-year old, it began:

> Here I am sitting at the library studying for a test on Tuesday in Mechanics. I'm hopelessly behind in the course and every five minutes I fall asleep for fifteen. Is this any way to spend a Sunday evening? What a crock of shit!

It then performed an initial stage of critical reflection:

> It seems that in the last years my ideas and values have changed greatly. I have always just taken things as they come, formed few plans for the future, and avoided any important

---

[19] I also arrogantly joined a senior-level English course on the 20th-century American novel, thinking I could handle anything on campus. I got my comeuppance quickly. Not only could I not understand the literary jargon in class discussions, I was also incapable of reading a book a week.

decisions. I am now in a process of soul-searching. What am I going to do with my life? ... I am really confused.

Although its desire was to connect passion with work, interest with beneficial contribution, it did not abandon socioeconomic prudence:

> I really feel like just getting up and saying forget it. I'm always asking myself what good is this or that course ever going to do me. This course is in my major so now I ask myself if I really want to be a mechanical engineer. It sounds so dull. When I get out of college do I really want to take some 9 to 5 job that offers a promising future financially but not much of anything else? But then again I might get into some dynamic position that would really interest me.

After a self-indulgent fantasy about just leaving (I was, and remain, enthralled with the Jefferson Airplane cover of "Wooden Ships"), it returned to an engineer's sense of efficiency:

> But ... why would any other place be any different than this? I would be traveling around looking for something that I can only find within myself. That would be a total waste.

I decided to expand my curricular travels beyond engineering without rejecting it and departing. I became the only student enrolled in the five-year arts-engineering option and began pursuing a B.A. in Social Relations (to my engineering student colleagues, pursuing a B.A. was beyond the pale, a waste of time). That meant engaging questions of social behavior more seriously. I especially enjoyed courses in cultural anthropology, which introduced me not only to a seemingly infinite range of different perspectives but also practices for investigating them. I became enthralled in particular with ethnography, studying others through interviews and participant observation with the goal of making visible meanings that were otherwise hidden.[20] For one course, I undertook an ethnographic study of life in my fraternity house called "House Behavior."[21] I approached my instructor, Barbara Frankel (Ph.D., Princeton), about possible graduate training in anthropology. She was enthusiastic. "You have to go to Chicago," she said. When I inquired why, she said, "Because it's the top program." It never occurred to me to ask who was on the faculty and whether or not their interests matched my own.

I anticipated the experience pursuing a Ph.D. in cultural anthropology largely in engineering terms, as an opportunity to acquire tools for analysis of human behavior.[22] Toward the end of my undergraduate years, I had turned to questioning "technology and society," not only because it linked my engineering past and present to a future in anthropology but also because it was a big question that transcended the particular. To an engineering professor who had agreed to write a letter of

---

[20] For an account of how education always conveys not only ideas but also specific work practices, see Sullivan and Rosen, *A New Agenda for Higher Education*, 2008. I participated in the seminar that yielded the book, and the *Engineering Cultures* course I discuss in my personal geography is profiled.

[21] Lehigh had thirty-five fraternities. Mine was for basketball players.

[22] For an analysis of engineering faculty during this period seeking to "appropriate" the social sciences for their own ends, see Wisnioski, "'Liberal Education Has Failed'," 2009.

recommendation, I explained that an earlier essay I had written "misleadingly overemphasizes my concern for environmental problems." My interest (like many others at that time) was "in the general interaction between man [sic] and his technological advancement."[23] To Dr. Frankel I wrote I was now "looking toward cultural anthropology for an introduction into the mysteries of man [sic; I had not read Marilyn French yet]." I was seeking to "supplement and complement" the education I had received so far. With that desire to make sure I "wasted nothing," I was just "reaching out to grasp a little more." Anthropology would enable me to understand "how and why reactions [to technological advancement] differ throughout the world" and how "we" can "begin to solve the multitudes of problems that are upon us as a result."[24] Just before leaving, I wrote a course paper on how the energy crisis might become a source for cultural change in the United States.

After joining the dual-degree program and gravitating toward anthropology, I also spent a summer in Europe to experience different perspectives firsthand. Even after working extra jobs and saving money for a year, I still needed support during the trip, so I found a job working on an archaeological excavation in the U.K. for $25 a week, including lunch. Traveling to Spain was the highlight, enabling me to finally live in Spanish for a short period. The language was a bit jarring, however, as I learned my dream had been Latin American. Latin American dialects of Spanish do not include the lisp.

Although I was taking a flying leap into a completely different academic world for graduate education, I did not realize just how much I was carrying my engineering identity and practices with me.[25] I was still engaging the world as a maze of problems to solve. I felt incomplete as a mechanical engineer because I did not gain mastery over the world of heat transfer. And while I critically analyzed the terms of narrow engineering service, I continued to carry the image of service as a desirable, even defining, work practice. During the summer before joining the graduate anthropology program at Chicago (which I selected again without visiting because it offered me a full scholarship), I worked as an engineer for a small firm that made glass bottles. It felt good to install the equipment that would help bring the plant into compliance with new air pollution regulations.

One hope I took with me to Chicago was that by learning to study controversies over technology I would find clarity and purity of purpose. In particular, by studying and making visible different stances on technology and society, might I find a place to stand myself? I ultimately found clarity of purpose, but at the cost of giving up on the comforts of purity.

MY SECOND CONSERVATISM: EXTERNAL CRITIQUE

I had no idea that the conceptual tools I would gain in becoming an anthropologist would generate as many problems for my career pathway as they resolved. At Chicago I was confronted immediately with my massive ignorance of intellectual traditions in Europe and the United States as well as a never-ending series of challenges to position myself in relation to those traditions. I had known that

---

[23]Downey, "Draft Letter from Gary Downey to Adrian Richards," 1973a.
[24]Downey, "Draft Letter from Gary Downey to Barbara Frankel," 1973b.
[25]For an account of the how an academic field "performs" through its practitioners, see Callon, "The Embeddedness of Economic Markets in Economics," 1998.

the anthropological concept of culture had emerged in the 1920s and 1930s and then made its way into popular discourse by the 1940s and 1950s, where it was still dominant. What I did not know was that Chicago stood for the superiority of analyzing cultures as systems or structures of meaning in opposition to both other academic approaches to understanding human action and the everyday image of cultures as patterns of beliefs and values.

Our many scholarly confrontations in courses, seminars, and discussions within the department did not challenge two key assumptions. One was that cultures were bounded wholes that overlapped in some way with collections of people (e.g., societies or countries). We focused debate on whether it was better to see these bounded wholes as "systems of symbols and meanings,"[26] "meaningful orders of people and things,"[27] or other candidate models. The other assumption was that shared cultures underlay the complex and situated interactions of everyday life. Culture lay "below." To that end, our studies of culture built upon an analogy with the grammatical structures of languages. Just as two people speaking a language shared an underlying grammar, we learned, so people interacting in everyday behavior exhibited the shared underlying meanings of culture.

I had two serious identity problems in this intellectual milieu. One was my interests lay not in shared underlying cultural grammars but in differences and conflict. Why were people fighting over nuclear power? How could the study of what combatants shared shed any light on their confrontations?

The other was I was interested in a material phenomenon: technology. I quickly lost count of the number of yawning responses I received from students and faculty to my expressions of interest in technology. In an intellectual world interested in shared ideas and deep underlying meanings, technologies were superficial phenomena, expressions of something else more fundamental. Practices of kinship and religion were certainly filled with meaning and fundamental to society, but technology? I had to prove, both to colleagues and to myself, that technology was worthy of study.

I occasionally considered initiating an ethnographic study of engineers. Despite all my talk about technology and society, my interest was never fundamentally in technologies *per se*. For me, the study of technology was a pathway toward understanding how people connected their knowledge to larger social and material projects (which I later came to understand as the normativities of knowledge in service).[28]

Each time, I postponed the idea of studying engineers on three grounds. The most obvious was that engineers appeared too diverse to share an underlying culture. From extensive personal experience in school and work, I understood innumerable contrasts among disciplines, positions in the workplace, relations among researchers and sponsors, etc., that were difficult to characterize as expressions of either one big culture or many little cultures. The second was that following engineers into the workplace also meant having to analyze the workplace itself. This necessarily involved wading into the complexities of political economy and the morass of debates over capitalism, far

[26] Schneider, American Kinship, a Cultural Account, 1968.
[27] Sahlins, Culture and Practical Reason, 1976.
[28] For discussion of the normativities of engineering service, see Brown et al., "Engineering Education and History of Technology," 2009.

from the relative safety of shared cultures. And third, I could not see how mapping everyday lives in the invisible worlds of engineering could speak in any way to the larger issues of the day. Debates over technology were capturing great popular attention. I took it as given that my effort to combine engineering with anthropology aimed at having service implications beyond the boundaries of the academy. At that time, routine engineering practices of design and manufacturing seemed absolutely marginal when compared to public controversies over technology (they are not).

I thus devoted much of my graduate career to developing strategies for making visible and analyzing meaning in everyday encounters with technology. My master's thesis was an anthropological view of the nuclear power controversy as a conflict in meaning. After years of anxious rumination, I came to see public conflicts over technologies as ideological conflicts. I followed the anthropologist Clifford Geertz in seeing ideologies not as perspectives that were flawed and opposed to factual accounts but as simply "schematic images of social order" that could simultaneously contrast with one another and be consistent with shared underlying culture.[29] Focusing on ideologies in conflict was a radical move in anthropology, for I saw ideologies as something people could choose or decide to accept.[30]

Throughout my graduate years, I continued my search for a broader social project from whose viewpoint all would become clear. I became enthralled with Marxist critiques of capitalism. There was much more to industry than I had seen on the job. I spent the better part of summer 1978 on a chaise lounge producing a fifty-odd page outline in engineer's microfiche handwriting of Marx's *Capital, Volume 1* (the volume he published before he died). His labor theory of value, which vested the ultimate value of goods in labor and characterized the capitalist workplace as a process of alienating workers from the products of their labor, made much sense to me. It helped me analyze hierarchies in the operations of companies I had earlier taken for granted. I also began to see connections between nuclear vendors and utility companies building custom-designed nuclear plants on the one side and supporting practices in the Nuclear Regulatory Commission and Congressional committees on science and technology on the other as key, constitutive relations of production.

I also experienced frustrations. What about technology in production? In Marx's works, I found only occasional references to technology (as both about relations of production and ideological superstructures) and little guidance. Also, what would life be like after a proletarian revolution? I had been persuaded, for example, by Crane Brinton's *The Anatomy of Revolution*, which held that the American and French political revolutions had produced as many continuities in hierarchical power as changes.[31] While inequalities generated through capitalist relations were certainly a severe problem, revolution would not necessarily eliminate or even reduce inequality.[32]

[29] Geertz, "Ideology as a Cultural System," 1973.

[30] My Ph.D. defense was nearly cut short when I responded "Yes" to a question from a faculty member who saw human behavior as always shaped by underlying social structures. "Are you saying," he asked, "people choose their perspectives in this debate?" I barely recovered by mapping linkages between the antinuclear ideology that was my focus and the presumably deeper cultural concept of the individual on which both sides of the debate depended. Resident historian of anthropology, George Stocking, later told me my dissertation defense "marked a turning point" in the department, the beginning of a shift in orientation to a "focus on agency." At the time, all I felt was pain.

[31] Brinton, The Anatomy of Revolution, 1952.

[32] I began wondering if engineers could introduce a sense of citizenship into capitalist accumulation working from the inside out.

I turned to nonviolent American anarchism, or what David DeLeon's helpful book *The American as Anarchist* called "left libertarianism."[33] In contrast to free-market libertarianism, which has received much attention since Reagan and Thatcher, left libertarianism is about building new social forms that would resist or sharply reduce hierarchy. I have written much about the attempts to abolish hierarchy and build alternative social relations among opponents of nuclear power and radioactive waste disposal. I was fascinated that the people who organized acts of nonviolent civil disobedience at nuclear power plants, most famously the Clamshell Alliance in New England in 1977, seemed to be shaping their lives and identities entirely around the analysis of and opposition to oppressive hierarchy. Also, since their actions routinely included small demonstrations of solar and other alternative technologies, opposition to nuclear power was not principled opposition to technology.[34] Here was purity of purpose, analysis, and technological practice! I was intensely jealous of their certainty. Their purity was achieved by positioning themselves external to nuclear power and the larger system it represented, and their work was dedicated to building alternative social forms.

I made studying them my dissertation topic, both to make their perspectives visible in an anthropological sense and to try them on myself. I followed antinuclear groups in Indiana, New York, and New England for two years. The more I attended meetings and conducted interviews, however, the more I found myself disaffected by a limitation that felt strangely familiar. The SHAD Alliance meetings (named after an endangered fish) I attended in Manhattan spent more time arguing over their discussion process (led by the rotating "vibes-watcher") than on analyzing and planning resistance to nuclear power plant construction. One weekend in Massachusetts, I watched 300 Clamshell Alliance members devote a forty-eight hour marathon to transforming their consensus decision-making process to a 90% majority rule so they could make any decisions on anything. Just as I had concluded after reading Marx that one cannot simply replace one society with another, I concluded that among antinuclear groups no pure critical stance or position external to dominant practices of power actually exists (including academic stances). One is always implicated (and hence complicit) in some way. Also, the internal pressure on members to subscribe to the left libertarian perspective ultimately felt to me as inflexible, controlling, and hence as conservative as my engineering curriculum and the expectation of white-collar work in private industry.

The issue of service for me became one of process rather than position—insuring that alternative practices stood a chance of challenging and replacing dominant practices (if such could be justified). Coming to this conclusion had a big implication for my personal geography. In order to be able to understand and affect dominant practices pertaining to technology, I had to locate myself in the midst of them (or in some cases simply accept my location within them). Through academic research, I could help make sure antinuclear groups (and others contesting specific technologies elsewhere) would not be invisible to centers of decision-making power.[35] Surely, any decision-making about nuclear technologies that did not take into consideration or explicitly repressed the views of

---

[33] DeLeon, The American as Anarchist, 1978.

[34] During this period, I learned from Thompson's *The Making of the English Working Class* that the early 19th century English Luddites were not Luddites either in the contemporary sense of anti-technology activists.

[35] I also ultimately became frustrated with the reverence in the liberal arts for isolated critical virtuosity. See concluding paragraph.

opponents (including both those committing civil disobedience and those working for administrative and legal remedies) was fundamentally flawed because it represented only particular interests.

Armed with an analysis of controversies as conflicts among ideological perspectives as well as a work practice making visible perspectives that were otherwise hidden, I became a student of technical controversies with an interest in seeking better outcomes by taking full account of ideological differences. One never escapes or eliminates conflict. Through my studies of the nuclear debate, for example, I had become an opponent. Participants in disagreements and conflicts, however, need not be entrenched opponents with incommensurable stances.

I decided to try government as a site for rethinking and redesigning processes of decision-making. In Spring 1980, the National Research Council (NRC), research arm of the National Academy of Sciences, awarded me a Mellon Foundation Postdoctoral Fellowship.[36] My assignment was to support and contribute to a newly-formed panel study of the social and economic dimensions of radioactive waste management.

I was indeed privileged to watch as staff from the Nuclear Regulatory Commission, Environmental Protection Agency, Congressional committees, and the Departments of Energy, Interior, and Defense came to testify about current states of knowledge in radioactive waste management. I came to see these meetings as analyzable and predictable confrontations of different perspectives both within and beyond the federal government. The committee itself was purposely diverse, including members from both a nuclear utility and Friends of the Earth. "I could make a difference here," I often thought. I imagined taking my engineering and anthropology into a career in science and technology policy, whether on Capitol Hill, in an agency, or representing an environmental group.

Yet this world had a big drawback: it had few engineers (I met only Jay Fay from MIT). Science and technology policy felt like a permanent step away from engineering. Becoming an expert on nuclear waste policy felt especially limiting, as I spent much time learning geo-hydrology. Where were the centers of power in engineering? What could I do to make visible perspectives that were otherwise hidden in engineering?

I figured a faculty position was my best pathway back to proximity to engineering. I had gained some experience teaching at a state college in Chicago, and the NRC study panel included distinguished professors who seemed to be successfully blending contributions inside and outside of their institutions. If I stayed in policy, I knew I would be unable to compete for an academic job after a few years. But after spending time in a university, D.C. would still be there. I had learned many things studying technology debates that could possibly benefit engineering students who may be learning practices that were self-limiting. I decided in 1981 to find out if education could be an effective site and vehicle for service that not only critiqued dominant practices but also made space for scaling up alternatives.

---

[36] My application had benefitted from the recommendation of Robert McCormick Adams, a Chicago archaeologist who was leaving the University to head the Smithsonian Institution. I learned of the National Academy of Engineering only after arrival, and was told by NAS President Phillip Handler it was mostly an honorary body that did not do much policy work.

CRITICAL PARTICIPATION AMONG ENGINEERS: MY PATHWAY TO *ENGINEERING CULTURES*

PEDAGOGY FOR ENGINEERING STUDENTS (BUT NOT ENGAGING THEIR IDENTITIES)

Teaching proved to be a good place for wrestling with dominant practices in engineering. At both Michigan Tech, where I taught for two years following my postdoc (1981-1983), and Virginia Tech, where I have taught since, I was hired to help build new programs in STS. STS was understood as Science, Technology, and Society if the focus was on making contributions beyond the academy and Science and Technology Studies if the focus was on producing a new type of research outcome. The variations and ambiguities have continued to this day (which I consider healthy). The Virginia Tech graduate program is called Science and Technology Studies but my department is called Science and Technology in Society.[37]

I had first encountered the term STS on my graduation day from Lehigh. As I was leaving the proceeding, I saw a poster announcing a new lecture series the following year on "Science, Technology, and Society." I stared in fascination and jealousy: "That's what I want to do." During my graduate career, the only reference to STS I could find was the sociologist Robert Merton's dissertation, *Science, Technology, and Society in Seventeenth Century England*, which I read closely for insights I could use.[38] By the early 1980s, STS was beginning to emerge as an interdisciplinary arena of scholarship and teaching. The Michigan Tech program was at the undergraduate level, and Virginia Tech was aiming to develop a Ph.D. program (which began in 1986).

The STS arena brought researchers together from history, sociology, and philosophy (not anthropology), as well as some from the natural and physical sciences.[39] Researchers found they needed to read and engage literatures beyond their fields of study in order to address problems within them. STS attracted financial support from universities in the form of faculty positions and new programs principally because science and technology had clearly become societal problems. The 1979 proposal for a new Center for the Study of Science in Society at Virginia Tech had focused on "science studies," defining it as "encompass[ing] all approaches to understanding the nature of scientific and technological activity as well as attempts to apply that understanding." Filled with references to science policy, science and technology as cultural institutions, and relevant activities of government and industry, it asserted that scientists "do not understand the nature of ideology, or their own psychology, their patterns of social behavior, or how their organizations work or what their policies are." There was no mention of engineering.[40] Joining the world of STS, the only survivable research identity I could adopt was to continue linking with technology. But I could engage engineers through teaching and outreach.

---

[37] I was hired into the Center for the Study of Science in Society. It later became the Center for Science and Technology Studies before ultimately becoming a department.

[38] Merton, "Science, Technology and Society in Seventeenth Century England," 1938.

[39] In 1987 and 1988, the anthropologist of science Sharon Traweek and I had session proposals on Anthropology of Science and Technology rejected by the American Anthropological Association on the grounds they fell outside the purview of anthropology.

[40] Bauer, "Science Studies (Science, Technology, and Society)," 1979.

Over the next fourteen years, I adopted four different approaches to helping undergraduate engineering students (and to a lesser extent students from other majors) critically examine their knowledge and commitments. Three of these were for engineers, but not specifically about them.

The first was congruent with the methodological strategies I had learned in anthropology: helping students make visible and analyze the common assumptions among them. In this case, the assumption was that technology was an autonomous force impacting on society. Engineering students tended to favor the development of new technologies, on the assumption technology has mostly positive benefits. My several versions of *Introduction to Humanities, Science, and Technology* examined emerging challenges to this assumption. The goal was to help engineering students recognize how making this assumption could have the effect of hiding from them responsibilities for the technological outcomes of their work. My teaching in this course drew heavily from the *Introduction to Cultural Anthropology* courses I was also teaching, which helped students make visible and critically analyze a variety of shared assumptions.

The second type of course further extended my anthropological strategies in an STS environment by challenging students to examine their assumptions about hierarchies in the world. The courses *Technological Change in Developing Countries* and *The West and the Rest* confronted students with substantial learning about technological and economic life beyond the United States. The goal was to make visible and critique common assumptions about international development and the superiority of the West. The second, for example, closely examined Sylvia Hewlett's *The Cruel Dilemmas of Development: Twentieth-Century Brazil* to explore structures of late capitalism and what Hewlett criticized as the "optimistic perspective" about "growth, freedom, and social justice."[41]

The third type extended my dissertation work analyzing the ideological dimensions of technical controversies (where people have the agency to choose perspectives) by making visible a variety of non-technical dimensions of technological controversies. Engineering students were learning to view their work as fundamentally technical in content. By helping them recognize and understand the social, cultural, and ethical complexities involved in technical work, I hoped they would be able to see that to focus only on technical contents was to willfully limit one's vision. I also hoped they would begin to analyze their own pathways and identities as having both technical and nontechnical contents.

I designed, for example, a three-course sequence *Science and Technology in Modern Society* to help students become more sophisticated decision makers (10-week courses on the quarter system). *Institutions in Science and Technology* examined who was involved in the development of science and technology, from the activities of federal science to industrial research and development (science-based). *Values and Value Conflicts in Science and Technology* explored how images of science and technology as vehicles of social progress emerged over time, and what was at stake in contemporary controversies. Finally, *Decisions and Decision Making in Science and Technology* gave students an opportunity to practice engaging conflict. Students in the 1986 version, for example, faced an in-class essay exam that located them as a knowledgeable "government official who frequently makes

---

[41] Downey, "Mid-Term Exam," 1984; Hewlett, *The Cruel Dilemmas of Development*, 1980.

decisions that involve science and technology, such as setting regulatory standards, siting hazardous facilities, and establishing criteria about new technologies." Although this official often made "explicit technical judgments," the judgments were "usually complicated" by "significant levels of uncertainty" and "sharp disagreement" among "interested groups." The challenge in the exam was to move beyond their usual approach of relying on "elite experts" and consider three alternative processes, including "making use of extensive public participation," "relying upon risk/benefit analysis," and using either "formal mediation" or the "science court." Most importantly, they had to take care to identify both the technical and nontechnical issues at stake.[42]

ENGAGING ENGINEERING IDENTITIES (IN BOTH TEACHING AND RESEARCH)

Engineering students clearly appeared to find these courses instructive and, sometimes, even liberating. Most had not previously had an opportunity to reflect on their assumptions about new technologies, international development, or the technical/social boundary in engineering education. Many accepted my challenges to become more sophisticated at positioning themselves in relation to public controversies over technology. Classroom discussions were energetic and students readily accepted the role-playing exercises I assigned and frequently led (e.g., create a debate over some emerging technology, or adopt a position other than your own and defend it). They gave me high student evaluations and I began winning teaching awards.

But what good were these experiences specifically for engineers? None dealt explicitly with engineers *qua* engineers. Reflecting the available literature and my personal experience, the courses on public controversies tended to assume government as the main site of decision making. Science and technology policy-making had few engineers. I pictured the vast majority of engineering students heading off to industry to solve problems they would expect others to define for them. They would likely understand ascending into management as departing engineering rather than realizing engineering in new venues. I worried these courses were providing engineering students with little more than an enjoyable respite from their engineering work, giving them little to take back to engineering curricula save a new awareness of complexity. How, I often wondered, would my students integrate an awareness of complexity into the next thermodynamics problem, the open-ended design course, or collaborative work on the job? Could I do more than treat them as STS apprentices, mini-graduate students, passing through my field as vacationing tourists?

I felt I bore a responsibility to build a program of both teaching and research on engineers and engineering identities. For this reason, I developed a fourth type of course specifically for engineers. I first taught *Development of the American Engineer* in 1983 at Michigan Tech.[43] The historian Terry Reynolds had designed it as a straight history course. In my hands, it became a course about the emergent identities of contemporary engineers in the United States. Engineering students often

---

[42]Downey, "Exam #2:LASc 3103 *Decisions and Decision Making in Science and Technology*," 1986a. Other courses that focused on making visible the nontechnical dimensions of technologies included *Nuclear Power and Public Policy*, *Nuclear Power in Society*, *Development of Radioactive Waste Management*, *Technologies under Fire*, and *Computers and Cultural Values*.
[43]Thank you, Larry Lankton.

observe that the purpose of studying history is to avoid the mistakes of the past. I challenged them to find the mistakes of the past (usually long buried) and suggested that questioning history might be helpful for better understanding their identities in the present. Reading extensively on the history of U.S. engineers, I developed a rudimentary teaching model of "descent lines" in engineering fields. It illustrated how civil engineers were connected to mining engineers and, much later, environmental engineers and how mechanical engineers emerged from manufacturing shops to beget virtually all other engineering fields, beginning with electrical, chemical, and industrial engineers.

At Virginia Tech, I initially taught the course under that title before getting it formally approved in 1988 as *The Technologist in Society*.[44] Said the course description during this period: "At first an aristocratic Lone Ranger who rode into town to harness nature with his technical expertise, the engineer later became a university-educated applied scientist and then an upper middle-class corporate employee." The purpose of the course was to "acquaint . . . students with the value-laden issues that have surrounded engineers and other technologists by examining their changing cultural identities in American society."[45] Much of the course was prelude to examining more closely the challenges of what I called the "Corporate Technologist."[46] Exams asked students to explain the reluctance of engineers to form unions, the ambiguities of engineering professionalism given their locations within corporations, how professional societies emerged, and what it might mean to argue that "characteristics of the culture of engineering are gender-related."[47]

I realized early on that a good way to help students better understand their identities and commitments as specifically American engineers was to incorporate material on engineers in other countries. I had regularly relied on transnational comparison in teaching anthropology and my other STS courses. In 1986, I began reading extensively about the history of engineers in other countries, especially in Europe and Japan, and worked comparative accounts into classroom discussions. An optional assignment in 1987, for example, asked students to "examine the cultural identities of four engineers who lived in different times and places, including at least one European engineer who was active prior to 1825." Students were to map engineering identities by examining the challenges engineers faced: "What did it mean to become an engineer? How did one become an engineer? . . . What kinds of conflicts did these individuals confront?"[48]

In 1988 and 1989, I took the step that gave me the idea and some of the practices for a greatly modified version of the course: I added a module on engineers in Japan (one week the first year; two weeks the second).[49] As I developed this module over time, it came to draw heavily from the work

[44] The permanent title did not refer specifically to engineers in order to make room for faculty who might be interested in teaching about other types of technologists.

[45] Downey, "The Technologist in Society," 1988a.

[46] To explore the challenges engineers felt in the corporate world, I relied initially on Peters and Waterman, *In Search of Excellence*, 1984. I later changed to Kunda, *Engineering Culture*, 1992. In January 1993, I enrolled in a training course in total quality management (TQM), sponsored by IBM. I wanted to understand how TQM used the idea of a "strong culture" to achieve outcomes that included empowering the bottom, downsizing the middle (including engineering design), and concentrating power at the top. Kunda's accessible study of the subject made it a superior replacement for Peters and Waterman.

[47] Downey, "Final Exam," 1986b.

[48] Downey, "Optional Project," 1987.

[49] Downey, "Syllabus," 1989.

of anthropologist Dorinne Kondo, who called attention to contrasts between the Western concept of the autonomous individual and the Japanese concept of persons defined in terms of their position in the household.[50] The success of this module in helping students recognize their own identities by comparing them to others led me to imagine a wholly modularized course built significantly around the emergence of engineers in different countries. Perhaps systematically studying engineers who valued different types of knowledge, occupied different types of positions, and exhibited different aspirations would make it easier for engineering students to analyze their own identities. I intensified the work examining the emergence of engineers and engineering in different countries, focusing on Britain, France, Germany, Japan, and the Soviet Union but also collecting materials on engineers everywhere.

Researching the history of engineers and engineering was part of my larger plan to build an arena of STS research dealing specifically with engineering and that would have meaning for engineers (now called Engineering Studies). While pursuing tenure and promotion by publishing my work on public controversies over nuclear technologies, I initiated outreach and research activities pertaining to engineers and engineering. Given the publication requirements of tenure, it was certainly foolish to persuade the dean of engineering to fund a series of lectures on Engineering in Society, with me as all-purpose organizer, beginning my first semester. At the same time, I was insistent that achieving tenure was not worth my time if it did not include using STS as a platform for challenging engineers to critically examine and engage their own identities.[51]

To build a program in Engineering Studies, I needed first to understand how and why STS researchers were not interested in engineering in the first place. Beginning in 1985, I supervised two undergraduate research projects on the topic "Humanities and Social Sciences Studies of Engineering." The plan was to "review recent research in these disciplines on the nature of engineering," including the "types of research that have been done and the types of questions asked."[52] Our data collection effort generated so much material that the question evolved into "How Engineering Ceased to Be a Problem in Science and Technology Studies," as the 1989 publication was subtitled.[53] We found that engineering had previously been a site of significant questioning in sociology, philosophy, and history, but the work had declined for reasons specific to each discipline.

In order to transition from the study of technology to the study of engineering, I initiated a project in 1987 examining the emergence of computer-aided design and manufacturing (CAD/CAM) within mechanical engineering. I had been watching throughout the 1980s how engineering and popular news publications portrayed CAD/CAM as a solution to national economic challenges from Japan and other countries by integrating design and manufacturing practices (e.g.,

---

[50] Kondo, *Crafting Selves*, 1990.

[51] I was (and am) not interested in producing anthropological/STS studies that limit themselves to seeking isolated critical virtuosity, achieving influence through diffusion alone, and accepting the comforts of resolute pessimism. See Downey, "What Is Engineering Studies For?" 2009.

[52] Elliott and Murdock, "Undergraduate Research Request Form: LASc 4990: *Humanities and Social Sciences Studies of Engineering*, 1985.

[53] Downey et al., "The Invisible Engineer," 1989. I included one of the students, Timothy Elliott, as co-author. I have throughout my career included graduate students as co-authors of publications provided they contribute at least some research to the project.

some predicted manufacturing could be fully automated by 2000). Studying CAD/CAM advocates produced a revelation that carried over to my teaching: engineers linked their identities to the identities of countries. It also sharpened the focus of my reading about engineers in other countries, looking for connections between the identities of the two.[54]

Gaining insight on how engineers valued different types of knowledge in different countries, I decided to initiate a focused ethnographic study of engineering education in the United States. I wanted to examine more rigorously how the core of engineering curricula, especially the engineering sciences, might be linked to the practices and identities of engineers at work. I initiated the project in 1992 with financial support from the National Science Foundation and research assistance from two new STS graduate students, Shannon Hegg and Juan Lucena.[55] My guiding hypothesis was that "the high value engineering curricula place on the calculus-based engineering sciences as the foundation of engineering problem solving is linked to the establishment of other social dimensions of engineering identities, and to the underrepresentation of women and minority students."[56] Sitting in engineering classes again made it clear to me that, in learning to solve engineering problems, students were learning to not engage (or even anticipate the existence of) others who defined problems and understood their work differently, including both engineers and non-engineers. As such, they were learning a narrow, self-limiting form of service that accepted a kind of servility. Findings from this project documented the sharp demarcation of learning practices in engineering from other curricula, dominance of problem solving in the engineering sciences, how learning problem-solving posed evolving challenges to the identities of students, the relative marginality of design, and the expectations that engineering work would be done mainly in private industry.[57]

INTEGRATING THE QUESTIONS OF LOCATION, KNOWLEDGE, AND DESIRE

My comparative research began to bear significant fruit when it occurred to me that engineers in different countries appeared to be engaging and responding to very different understandings of human progress. I had long been teaching about the emergence of an emphasis in the United States (especially after the Civil War) on low-cost production in the private sector for mass production. I was greatly struck by repeated references in the French literature to the ideal and possibility of achieving perfection. Mass production was never about achieving perfection. The German literature made frequent reference to the idea of emancipating spirit. The British literature was tricky, for civil engineers were prominent in the early 19th century but emergent fields seemed to be stuck by their proximity to manual labor. The Soviet Union was, of course, never about private sector production,

---

[54]For accounts of the national narrative of economic competitiveness in engineering work, see Downey, "Agency in CAD/CAM Technology," 1992b; Downey, "CAD/CAM Saves the Nation?" 1992c; Downey, *The Machine in Me*, 1998b, Chapter 1.

[55]Juan Lucena in particular embraced the work of this project, conducting many interviews and observing student events.

[56]Downey, "NSF Proposal," 1992a, 1.

[57]Downey and Lucena, "Engineering Selves," 1997b; Downey and Lucena, "When Students Resist," 2003. An especially key moment for me came in 1992 after I presented initial findings from this project to a collection of anthropologists I and Joseph Dumit had gathered at the School of American Research, an anthropology think tank. In conversation, Paul Rabinow asked, "But Gary, what is engineering for?" Pursuing that question came to define my work ever since.

and prior to Stalin's ascent in 1928 it was certainly about achieving an ideal of equality.[58] Realizing these differences gave me the organizing theme for an anthropologized version of the course, one that would compare the emergence of engineers and engineering identities in relation to different concepts of progress.

To be truly compelling to engineering students, this revision would also have to go past my other courses and offer them specific practices they could take back to their curricula. It would have to be fully learner-oriented. The practices I offered would have to make clear the limitations of engineering problem solving without ever appearing to dismiss or reject its well-established and highly valued practices (to fulfill my commitment to critical participation). I found a solution in the words "location," "knowledge," and "desire." My research was demonstrating that engineers were located differently in different countries and valued different types of knowledge. Also, all individual engineers and non-engineers had unique life histories, and so came to work with different configurations of desires. Since engineering problems never solve themselves but are always solved by people, engineering work necessarily involved engaging people who defined problems differently, understood themselves and their work differently, and carried desires for different types of outcomes.

A revised version of *The Technologist in Society* could potentially ingratiate itself deep into the learning of engineers by helping students learn how to question their own locations, forms of knowledge, and desires in relation to those who answered these questions differently. The core engineering science curriculum focused plenty on engineering problem solving but not on engaging people who defined problems in different ways. How are engineers located, in relation both to one another and to non-engineers? What do they know? What do they want? Helping students gain the ability to ask these questions intelligently could make this elective an integral component of engineering education. By 1994, I had acquired sufficient knowledge about the emergence of engineers in five other countries to prepare detailed lecture material. It was time to try out a version that would share this knowledge with students while training them in the practices of questioning location, knowledge, and desire. I collected the questions together under the label "Mapping Engineering Problems through Humans."[59]

The practices of mapping the locations (including the power dimensions of location), forms of knowledge, and desires of stakeholders in engineering decisions were by no means a solution to overt conflict. Minimally, the benefit to students would be to inhibit a common tendency to judge those who held radically different perspectives, especially non-engineers, as simply lacking in knowledge or rationality. Making visible the knowledge dimensions of conflicts could indeed open new pathways for resolving them, even if not necessarily so. The great value in inhibiting knee-jerk rejections would not be that it reduced conflict but that it made clear the identities of both engineers and non-engineers are always at stake in the technical judgments of engineers. It would also make

---

[58] Graham, The Ghost of the Executed Engineer, 1993.

[59] The final exam for the second version of the course included the question: "Gary Downey's approach to Mapping Engineering Problems through Humans involves asking questions in the three categories he calls _____, _____, and _____. [note: knowing the exact words is not necessary, but try to approximate the concepts.]" The exact words were location, knowledge, and desire.

clear that the work of engineers always has nontechnical dimensions and includes commitments to broader social projects.

For engineering students who were in the process of solving hundreds of problem sets each of which began with the word "given," questioning location, knowledge, and desire could give them transportable practices for examining their own knowledge and commitments and engaging in more knowledgeable and sophisticated ways others who defined problems and understood their work differently. What they would need to take away was the mnemonic trigger: "location, knowledge, and desire."

I ruminated for a long time before deciding to change the course title to *Engineering Cultures*. Anthropologists were abandoning the culture concept because it tended to enforce an image of geographically bounded wholes (anachronistic in an increasingly networked world) and an internal homogeneity that erased hierarchy and difference.[60] But I had to consider the audience. To engineers, the word culture signaled important considerations in engineering work that varied around the world.[61] I decided to use the title to draw students in and then teach them an approach to cultural analysis that focused on dominant images and practices rather than bounded and homogeneous cultures.[62]

Introducing engineering students to an elective course focused on international education was thus not a means to increase their competitiveness for jobs nor to help the United States increase its economic competitiveness.[63] It was also not about global citizenship, sustainability, or exploitation. It was rather a strategy to help students become critical analysts of their own knowledge, commitments, and trajectories, which I judge to be the necessary first step in reflecting on any of the normativities of engineering formation and practice.

My formal course proposal in 1995 promised understanding of the "[d]evelopment of engineering and its cultural roles in historical and cross-national perspectives." Students would emerge understanding the "roles of engineers and engineering in popular life, development of national styles,[64] changing values in engineering problem solving, and effects of evolving forms of capital-

---

[60]Clifford et al., *Writing Culture*, 1986; Marcus and Fischer, *Anthropology as Cultural Critique*, 1999.

[61]I shared all my plans for revising *The Technologist in Society* with Juan Lucena, who served as my teaching assistant during the first three semesters of the *Engineering Cultures* course. He contributed some helpful content to the module on France from his Enlightenment course with Professor Ann Laberge. His prior decision to not simply inhabit the elite spaces in which he had been raised in Colombia and the United States both captured my heart and demonstrated the commitment to critical self-analysis necessary to help students do the same. In the second semester, he led a class on engineers in Colombia and in the third developed and taught a module called "Latin American perspectives." After finishing his dissertation and leaving in 1996, he expanded this module into the important new course *Engineering Cultures in the Developing World* (Brazil, Colombia, and Mexico). Between 1998 and 2006, he joined me in delivering several presentations about *Engineering Cultures*, and we sometimes alerted one another about new source material. He initially suggested introducing *Engineering Cultures* to the American Society for Engineering Education in 1998, and he arranged for us to teach a two-week version jointly in Paris in 2001. See his chapter for an account of his struggles integrating the course into liberal arts curricula at two institutions and concerns he has about the contents of his version.

[62]In this approach, dominant practices always stand in relation to subordinate practices. For a more theoretical account of dominant practices, see Downey, "What Is Engineering Studies For?" 2009.

[63]But as I indicated above, at least some students take away nothing more.

[64]The course never actually used the problematic term "national style," which invokes an image of coherent personhood. I included it in the proposal knowing curriculum committees would find it clean and, hence, appealing.

ism." They would be able to compare dominant practices in the United States with those in at least six other countries. To appeal to engineering faculty, the course did promise to improve students' abilities "to function effectively as engineers in a globalizing world" (by which I meant to always examine and search for ways of accommodating the location, knowledge, and desire of any stakeholder).[65]

SCALING UP: REDEFINING ENGINEERING AND BUILDING A NEW FIELD OF SCHOLARSHIP

Be careful what you ask for; you just might get it. Four years after first offering *Engineering Cultures*, I sold my soul to distance education. The course had rapidly grown far more popular than I could handle (from twenty-two in 1995 to more than 250 requests (150 accepted) three semesters later). To accommodate the demand and insure a personalized learning experience for each student (essential to developing practices of critical self-analysis), I reorganized classroom pedagogy by introducing discussion sections with graduate assistants who supervised the many homework and classroom exercises in manageable groups of twenty-five students each. An additional goal was to mentor assistants to teach their own versions of the course, both within and beyond the University.[66] But I was turning many students away, and I was training graduate assistants who could in principle manage online courses if I provided the content. I felt compelled to take advantage of a Virginia Tech initiative to position itself on the forefront of instructional technology. The new Center for Innovation in Learning awarded me the first of three substantial grants to develop a multimedia version of the *Engineering Cultures* modules.

My idea in 1999 was to produce interactive multimedia experiences that would facilitate student learning through questions and answers, parallel to what took place in classroom discussions. I hired three STS graduate assistants to initiate the process.[67] They would assist me in breaking down thirty-nine hours of transcribed audio recordings from the classroom course (from a semester with no discussion sections), translating these into corresponding multimedia presentations, and then reassembling them into clear, easy-to-use CDs. The proposal promised class lectures on short video clips (five minutes) and longer audio files (thirty minutes) as well as detailed notes and commentary on readings, reviews of relevant bibliographies, and a multidimensional video timeline that would compare what was going on with engineers in different countries at similar points in history.

[65] Downey, "Course Proposal," 1995. To appeal to students who disliked enrolling in any elective courses, I secured approval to cross-list the course in both the STS and History curricula and to have it count toward two areas of the core curriculum (much dreaded for many students).

[66] To date I have mentored twenty-four present and former graduate students and one colleague in Science and Technology Studies to teach the course. The majority of the graduate students have made substantial contributions to its curriculum and eleven have taught their own versions [indicated by *]: Donna Augustine, Frankie Bausch, Thomas Bigley*, Ben Cohen, Sharon Elber*, Thomas Faigle*, Wyatt Galusky, Krista Gile, Chris Hays, Grace Hood, Brent Jesiek*, Theresa Jurotich, Liam Kelly*, Gouk Tae Kim*, Jongmin Lee, Jane Lehr*, Juan Lucena*, Jonson Miller*, Amy Nichols-Belo, Robert Olivo*, Sunita Raina, Nicholas Sakellariou, Deanna Spraker*, and Lai-ju Zhang. My colleague Matthew Wisnioski began teaching his own version in Spring 2009, mentoring students as well.

[67] Wyatt Galusky, Heather Harris, and Jane Lehr.

The plan was hubris, born of naiveté. Audio transcriptions of classes did not count as content in this new medium. Producing just minutes of interactive multimedia material would take months of work, and I had committed to teach a 100% online version of the course in just over a year.

The content for CDs became multimedia lectures. I wrote and delivered twenty-eight lectures totaling thirty-six hours. Each time I had to sit motionless in a chair (to minimize file size) while endeavoring to wax enthusiastic. My teaching assistants regularly used the CD presentations that resulted (talking head with supporting text and images) in online versions of the course. I soon found myself wondering if these were functioning well locally because I was a known and accepted local authority. Colleagues at other institutions preparing students for international experiences were finding the material less attractive. One told me her students would prefer to engage the material in a book than watch and listen to me on a computer screen. Would sending students to multimedia CDs produced by an outside "expert" amount to giving up control of one's curriculum?[68] Nevertheless, in 2008 I worked with programmers at two institutions to make nineteen of these presentations available for free at www.globalhub.org.[69] The next step is indeed to convert them to written form and publish them as short textbooks.

By 1998, my syllabus had become a relatively coherent narrative transporting students from unreflective acceptance of everything they encountered in engineering to an ability to critically analyze their positions in relation to others (when it worked well). Several teaching assistants took initiative to devise new exercises and simulations that mimicked confrontations among engineers and others located differently from one another and with different forms of knowledge and desire. I systematically replaced the word "culture" with "dominant images," encouraging students to begin recognizing themselves as responding to a variety of dominant images that challenge them.[70] What was shared in different countries was not underlying meanings but challenges from overarching dominant images, to which individuals respond differently from one another. And perhaps most importantly, I introduced the Critical Reflections Assignments as a vehicle for helping students assess the development of their own critical awareness and practices. These became stabilized in an initial essay assignment asking students to map their trajectories into engineering and a final essay assignment asking them to "map who you are now having taken the course."[71]

By 2000, students in the course were able to invoke dominant images of progress to explain, for example, a sharp contrast between the high status of the most elite engineers in France and the

---

[68] In 2007, a leadership team at Michelin Americas Research Center (Greenville, South Carolina) drew on the France module and an on-site workshop to bring their work practices more in line with their senior colleagues in France. An excited member of the team later called me to say, "It worked; they loved it." What he meant was his team got the new tire they wanted by proposing to the home office in Paris not a new tire but a new mathematical "research method" for tire design. What I did not hear, however, was evidence of significant critical self-analysis on the American side.

[69] Early proposals to Pearson Publishing and McGraw-Hill to publish the multimedia presentations themselves had proved unsuccessful.

[70] Dominant images are simply what one takes as given. See Chapter 1 of Downey, *The Machine in Me*, 1998b. For a more recent theoretical account of dominant images and practices, see Downey, "What Is Engineering Studies For?" 2009.

[71] An unpublished content analysis of 120 final "critical reflections" (scored by an STS graduate student not involved in the course) in 1998 categorized eighty-six (72%) as "strongly positive," thirteen (11%) as "positive," thirteen (11%) as "mixed," and nine (7%) as "negative." He reported that in many cases the strongly positive assessments "indicated that students' experiences in the course transformed them in a profound way."

two centuries of struggles by British engineers at all levels to be seen as significantly distinct from manual labor. Students had to demonstrate some understanding of how elite engineers in Germany successfully gained acceptance for the view that *Techniks* (technical work and its products) could be key sites for the emancipation of *Geist* (human mind/spirit). Students sometimes had to write poetry illustrating Soviet dreams for a world celebrating technical work and free of inequalities produced through capitalism. They sometimes had to explain why some Japanese engineers began only in the past decade to assert professional identities apart from positions within large corporate *ie* (households; pronounced ee-aa [quickly, not to be confused with the Japanese word for "no"]). And they had to explain how the vast majority of U.S. engineers came to expect their careers to be spent in the private sector producing low-cost goods for mass consumption.

Interest in global education for the purposes of competitiveness expanded greatly after publication in 2000 of new accreditation criteria for engineering programs (Criterion 3h: "ability to understand the impact of engineering solutions in a global context") and, especially, Thomas Friedman's *The World is Flat* in 2005. This was not the same as integrating practices of critical self-analysis. My first reaction was to abandon the image of the "global" entirely. My second reaction was to call attention to what had emerged in my work as the key content of global competency, the ability to work effectively with people who define problems differently, including non-engineers.

I organized a research project with eight others to develop a criterion and three learning outcomes for global competency. We then developed and tested three instruments for measuring student progress toward these outcomes, including one wholly new type. It was a pre/post essay designed to test students' abilities to "analyze how people's lives and experiences in other countries may shape or affect what they consider to be at stake in engineering work."[72] While the results were encouraging, I have not yet conducted a longitudinal study testing the longevity of questioning location, knowledge, and desire as an engineering practice.

My curricular work began to step beyond *Engineering Cultures* and dive deeper into the heart of engineering curricula in 2004 when I was invited to deliver a keynote address to the World Congress of Chemical Engineering on engineering education. In preparing this address, it suddenly occurred to me that while the practices of collaborative problem definition emphasized in my course may be supplemental to practices of mathematical problem solving, they were not supplemental to everyday engineering work. Indeed, problem definition was both an essential part of engineering practice and upstream of problem solving. I decided to contest the core of engineering curricula, arguing that core practices should include collaborative problem definition alongside mathematical problem solving. The address elaborated a well-known design case in chemical engineering to outline practices integrating problem definition. These included analyzing how the engineering sciences are taught as separate, limited worlds; integrating technical electives focused on difficulties in collaborative problem definition; and introducing curricular tracks within existing fields focused on engineering design, engineering and management, engineering and policy, and engineering and society.[73] I now

---

[72]Downey et al., "The Globally Competent Engineer," 2006.
[73]Downey, "Keynote Address," 2005.

fear I am giving myself the task of someday having to write a textbook on engineering statics built on an image of engineering as problem definition and solution (PDS). The likely next step is case studies of conflicts in engineering problem definition to use in technical electives.

Many researchers have written about the broader social projects served by engineering practices. These range from linkages to capitalism and colonialism to service to domestic government bureaucracies and affirmation of dominant gender categories. My own research examining how dominant practices of engineering education and training tend to respond to variable and changing metrics of material progress, combined with my awareness and appreciation for the work of others, has grounded my efforts to build a field of scholarship devoted to examining the knowledge and normativities of engineering practices. I envision this field—called Engineering Studies—as not only analyzing but also participating critically in practices of engineering training and work.

At this writing, I am devoting much of my time to building Engineering Studies (this book is an example). I am driven by the importance of calling explicit and regular attention to the vast diversity of ways in which millions of engineers have served broader social projects in developing and enacting forms of knowledge and associated practices. Such work is essential, I maintain, to help ensure practicing engineers critically examine their knowledge and commitments in relation to those projects and take more active responsibility for the contents and effects of engineering service.

In 2002, I organized a meeting of twenty researchers in STS and history of technology with the goal of establishing an organization devoted to Engineering Studies. The historian Maria Paula Diogo (University of Lisbon) and sociologist Chyuan-Yuan Wu (National Tsing Hua University, Taiwan) joined me in establishing the International Network for Engineering Studies (INES) in Paris in 2004 and participated in its first workshop held at Virginia Tech in 2006.[74] In 2007, Juan Lucena joined me as co-editor of the organization's new journal *Engineering Studies*.[75] In 2009 and 2010, Atsushi Akera (Rensselaer Polytechnic Institute) led INES workshops in Lisbon and Troy, New York, respectively. In 2010, I established the *Engineering Studies Series* with MIT Press to serve as the intellectual capstone of the field and the *Global Engineering* series with Morgan & Claypool Publishers (with Kacey Beddoes as Assistant Editor) to provide instructional materials to engineering students and working engineers in the form of short books.

The prospects for greatly expanding Engineering Studies and its commitment to analyzing and contributing to engineering knowledge in service seem promising. Engineering fields have been characterized too often by an absence of self-criticism, the liberal arts by reverence for isolated critical virtuosity and acceptance of the comforts of resolute pessimism. My hope in building Engineering Studies as an international field of teaching, research, and outreach is to blur the boundaries between them by making clear that normative commitments in engineering work permeate it with liberal arts contents. I work with others to call greater attention to the variable, overlapping, and

---

[74]www.inesweb.org.
[75]www.tandf.co.uk/journals/engineeringstudies.

frequently competing commitments of engineering knowledge and practices. My ultimate goal is to help overcome self-limitations in engineering service by strengthening it.

REFERENCES

Alder, Ken. *Engineering the Revolution: Arms and Enlightenment in France, 1763–1815.* Princeton, NJ: Princeton University Press, 1997. 389

Bauer, Henry. "Science Studies (Science, Technology, and Society): A Proposed Center in the College of Arts and Sciences." Blacksburg, VA: Virginia Tech, 1979. 398

Beer, Ferdinand P. and E. Russell Johnston Jr. *Mechanics for Engineers: Statics.* New York, NY: McGraw-Hill Book Company, 1962. 389

Brinton, Crane. *The Anatomy of Revolution.* New York, NY: Prentice-Hall, 1952. 395

Brown, John K., Gary Lee Downey, and Maria Paula Diogo. "Engineering Education and History of Technology." *Technology and Culture* 50, no. 4 (2009): 737–752. 394

Callon, Michel. "Introduction: The Embeddedness of Economic Markets in Economics." In *The Laws of the Markets*, edited by Michel Callon, 1–57. Oxford; Malden, MA: Blackwell Publishers/The Sociological Review, 1998. 393

Clifford, James, George E. Marcus, and School of American Research (Santa Fe, NM). *Writing Culture: The Poetics and Politics of Ethnography: A School of American Research Advanced Seminar.*Berkeley, CA: University of California Press, 1986. 405

Cole, Robert. "The Rights of Spring: A Forum Answers an Old Question of Power." *Lehigh University News*, 1970. 387

DeLeon, David. *The American as Anarchist: Reflections on Indigenous Radicalism.* Baltimore, MD: Johns Hopkins University Press, 1978. 396

Downey, Gary. "Draft Letter from Gary Downey to Adrian Richards." <manuscript>, 1973. 393

Downey, Gary. "Draft Letter from Gary Downey to Barbara Frankel." <manuscript>, 1973. 393

Downey, Gary. "Letter from Gary Downey to G.J. Huebner, Jr." <manuscript>, 1973. 390

Downey, Gary. "Mid-Term Exam: LASc 3850 *The West and the Rest.*" <manuscript>, 1984. 399

Downey, Gary. "Exam 2: LASc 3103 *Decisions and Decision Making in Science and Technology*<manuscript>, 1986. 400

Downey, Gary. "Final Exam: LASc 3850 Development of the American Engineer." <manuscript>, 1986. 401

Downey, Gary. "Optional Project: LASc 3850 *Development of the American Engineer.*" <manuscript>, 1987. 401

Downey, Gary. "LASc 3850 the Technologist in Society." <manuscript>, 1988. 401

Downey, Gary. "Syllabus: Hum 3854 *The Technologist in Society.*" <manuscript>, 1989. 401

Downey, Gary. "NSF Proposal: Constructing Engineers: A Participant-Observation Study of Undergraduate Engineering Education." <manuscript>, 1992. 403

Downey, Gary. "Course Proposal: Humanities, Science, and Technology 2054: Engineering Cultures." <manuscript>, 1995. 406

Downey, Gary Lee. "Agency in CAD/CAM Technology." *Anthropology Today*, October (1992): 6–10. 403

Downey, Gary Lee. "CAD/CAM Saves the Nation?: Toward an Anthropology of Technology." *Knowledge and Society* 9 (1992): 143–168. 403

Downey, Gary Lee. *The Machine in Me: An Anthropologist Sits among Computer Engineers.* New York, NY: Routledge, 1998. 389, 403, 407

Downey, Gary Lee. "Keynote Address: Are Engineers Losing Control of Technology?: From 'Problem Solving' to 'Problem Definition and Solution' in Engineering Education." *Chemical Engineering Research and Design* 83, no. A8 (2005): 1–12. 386, 408

Downey, Gary Lee. "The Engineering Cultures Syllabus as Formation Narrative: Critical Participation in Engineering Education through Problem Definition." *St. Thomas Law Journal (special symposium issue on professional identity in law, medicine, and engineering)* 5, no. 2 (2008): 101–130. 386

Downey, Gary Lee. "What Is Engineering Studies For?: Dominant Practices and Scalable Scholarship." *Engineering Studies: Journal of the International Network for Engineering Studies* 1, no. 1 (2009): 55–76. DOI: 10.1080/19378620902786499 386, 402, 405, 407

Downey, Gary Lee, Arthur Donovan, and Timothy J. Elliott. "The Invisible Engineer: How Engineering Ceased to Be a Problem in Science and Technology Studies." *Knowledge and Society* 8, (1989): 189–216. 402

Downey, Gary Lee and Joseph Dumit. "Locating and Intervening." In *Cyborgs and Citadels: Anthropological Interventions in Emerging Sciences and Technologies*, edited by Gary Lee Downey and Joseph Dumit, 5–30. Santa Fe, NM: The School of American Research Press, 1997.

Downey, Gary Lee and Juan C. Lucena. "Engineering Selves: Hiring in to a Contested Field of Education." In *Cyborgs and Citadels: Anthropological Interventions in Emerging Sciences and Technologies*, edited by Gary Lee Downey and Joseph Dumit, 117–142. Santa Fe, NM: The School of American Research Press, 1997. 386, 403

Downey, Gary Lee and Juan C. Lucena. "When Students Resist: Ethnography of a Senior Design Experience in Engineering." *International Journal of Engineering Education* 19, no. 1 (2003): 168–176. 403

Downey, Gary Lee, Juan C. Lucena, Barbara M. Moskal, Thomas Bigley, Chris Hays, Brent K. Jesiek, Liam Kelly, Jane L. Lehr, Jonson Miller, Amy Nichols-Belo, Sharon Ruff, and Rosamond Parkhurst. "The Globally Competent Engineer: Working Effectively with People Who Define Problems Differently." *Journal of Engineering Education* 95, no. 2 (2006): 101–122. 386, 408

Elliott, Timothy and Kelly Murdock, "Undergraduate Research Request Form: LASc 4990: *Humanities and Social Sciences Studies of Engineering*," <manuscript>, 1985. 402

Forum, Lehigh University. "Constitution of the Lehigh University Forum." Bethlehem, PA: Lehigh University, 1970. 387

Forum, Lehigh University. "Motions Passed by the Lehigh University Forum: Fall Semester 1970." Bethlehem, PA: Lehigh University, 1970. 388

French, Marilyn. *The Women's Room*. New York: Summit Books, 1977. 390

Geertz, Clifford. "Ideology as a Cultural System." In *The Interpretation of Cultures, Selected Essays*, 193–233. New York, NY: Basic Books, 1973. 395

Gerstl, Joel E. and Robert Perrucci. *Profession without Community*. New York, NY: Random House, 1969. 389

Graham, Loren R. *The Ghost of the Executed Engineer: Technology and the Fall of the Soviet Union*. Cambridge, MA: Harvard University Press, 1993. 404

Hewlett, Sylvia Ann. *The Cruel Dilemmas of Development: Twentieth-Century Brazil*. New York, NY: Basic Books, 1980. 399

Kondo, Dorinne K. *Crafting Selves: Power, Gender, and Discourses of Identity in a Japanese Workplace*. Chicago, IL: University of Chicago Press, 1990. 402

Kunda, Gideon. *Engineering Culture: Control and Commitment in a High-Tech Corporation*. Philadelphia, PA: Temple University Press, 1992. 401

Marcus, George E. and Michael M. J. Fischer. *Anthropology as Cultural Critique: An Experimental Moment in the Human Sciences*. Chicago, IL: University of Chicago Press, 1999. 405

Merton, Robert K. "Science, Technology and Society in Seventeenth Century England." *Osiris* 4, no. 1 (1938): 360–632. 398

Peters, Thomas J. and Robert H. Waterman. *In Search of Excellence: Lessons from America's Best-Run Companies.* New York, NY: Warner Books, 1984. 401

Richards, Adrian. "CE 332: *Ocean Engineering* Final Examination, Department of Civil Engineering, Lehigh University." <manuscript>, 1973. 390

Sahlins, Marshall David. *Culture and Practical Reason.* Chicago, IL: University of Chicago Press, 1976. 394

Schneider, David Murray. *American Kinship, a Cultural Account.* Englewood Cliffs, NJ: Prentice-Hall, 1968. 394

Sullivan, William M. and Matthew S. Rosen. *A New Agenda for Higher Education: Shaping a Life of the Mind for Practice.* San Francisco, CA: Jossey-Bass, 2008. 392

Thompson, E. P. *The Making of the English Working Class.* New York, NY: Vintage Books, 1966.

Wisnioski, Matthew. "Liberal Education Has Failed': Reading Like an Engineer in 1960s America." *Technology and Culture*, 50 (2009) 4: 753–782. 392

Yates, W. Ross. *Lehigh University: A History of Education in Engineering, Business, and the Human Condition.* Bethlehem, PA: Lehigh University Press, 1992. 388

Epilogue
Beyond Global Competence: Implications for Engineering Pedagogy

Gary Lee Downey

A map has no conclusion, yet it does warrant responses. This epilogue is one response to the map of trajectories into international and global engineering education that constitutes this volume. Drawing on the personal geographies, it addresses the question: What implications do these diverse trajectories have for dominant practices of engineering pedagogy?

A main source of resistance to expanded participation in international and global engineering education is widespread acceptance by engineering faculty of the view that an engineering graduate from a given field is at the core basically one thing. That one thing is technical competence in a collection of engineering sciences. The emphasis on competence in engineering sciences makes it difficult for engineering faculty to see and critically analyze how they also learn through engineering education to accept and follow broader social projects that rise to dominance, e.g., economic competitiveness. This view inhibits engineering curricula from placing value on other types of knowledge beyond the engineering sciences and on other types of external projects.

A key challenge to international and global engineering educators (as well as to other educators who seek to redefine what counts as engineering and what it means be an engineer) is to find ways of integrating into engineering curricula the practices of critical self-analysis that confronted them as they added identities outside home countries. Such depends upon examining how dominant practices in engineering curricula challenge students to add knowledge and normative commitments at the same time.

THE DOMINANT PRACTICES OF "CORE VS. PERIPHERY"

These personal geographies make it clear that engineering faculty (by and large) do not care about international and global engineering education. The many struggles contributors report indicate that engineering faculty generally do not judge education beyond the country to be necessary or integral to the formation of engineers. Nowhere does international and global education appear as a requirement in engineering curricula.[76] Engineering faculty, especially those teaching the engineering sciences, may even view efforts to incorporate international and global education within engineering as threatening the technical quality of existing engineering courses. Their own identities are at stake as the providers of quality engineering education.

The personal geographies also suggest that engineering students (by and large) care about international and global engineering as much as faculty do. Faculty show students what they judge to be important through curricular requirements, and students follow. It is difficult to expect a significant proportion of engineering students to seek experiences in international and global education if their faculty mentors do not consider it valuable.

For engineering faculty, a dominant curricular image and pedagogical practice involves separating core from periphery.[77] This separation establishes a clear hierarchy. The core is technical and important. It consists of required and elective courses in the engineering sciences in which the dominant practice is mathematical problem solving. The periphery is judged to introduce non-technical contents and hence is of lesser importance, sometimes no importance. Examples closer to the core are "introduction to engineering" experiences in the first year and "capstone design experiences" in the final year. Further away are non-technical electives in the humanities and social sciences. One cannot claim the identity of engineer without acquiring the core practices of engineering problem solving.

As an example, see Figure 18.1, the degree path sheet for Virginia Tech mechanical engineering students through the class of 2012.[78] The main flow is from top to bottom, first year to fourth year. The core of the curriculum is defined by the dense network of courses in the engineering sciences, indicated by the connected boxes (all you need be able to see are the connecting lines, or their absence). The curriculum has three core strands, including "mechanical, materials, & processes" in the center, "electrical, instrumentation, measurement" to the left, and "thermal fluid sciences" to the right. The core culminates in courses in senior design. Off to the right are the peripheral courses, connected to nothing. These are judged to be nontechnical in content. Courses in the humanities and social sciences fall there.

Actual engineering curricula differ across distinct fields and institutions. These differences warrant close attention, indeed demand it. At no institution in the United States granting degrees in engineering, however, do engineering faculty or students have to explain or justify a separation between technical core (with its practices of mathematical problem solving) and nontechnical periphery. It is a distinction whose acceptance has scaled up sufficiently to constitute dominance. It is a given. As a given, it becomes an elephant in the room, taking up enormous space yet difficult to see and name. It is also, as we shall see, not the only elephant. The acquisition of practices and

---

[76] See Les Gerhardt's chapter for an account of Rensselaer Polytechnic Institute initiating in 2009 a process designed to achieve an Institute-wide requirement for some form of international and global education outside the country.

[77] Downey, *The Machine in Me: An Anthropologist Sits among Computer Engineers*, 1998, 128–165; Downey and Lucena, "Engineering Selves: Hiring in to a Contested Field of Education," 1997.

[78] Department of Mechanical Engineering, Virginia Tech, B.S. In Mechanical Engineering: Proposed Degree Path Sheet - Classes of 2009, 2010, 2011, and 2012, 2009.

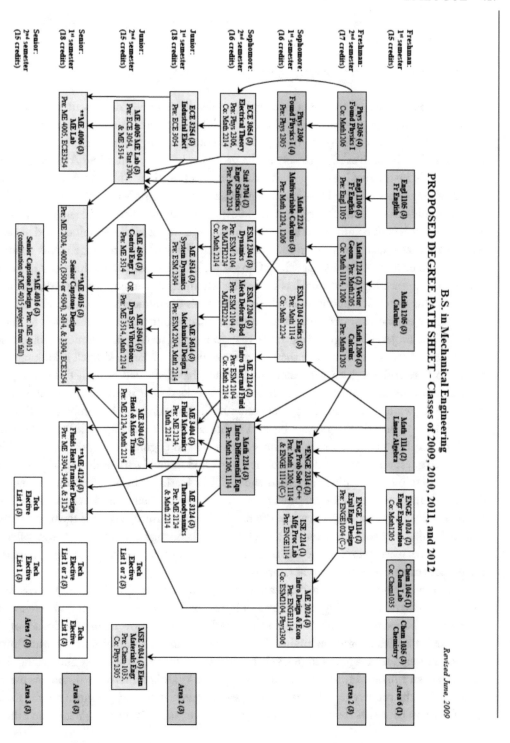

Figure 18.1: A Degree–Path (Faculty) View of the Engineering Curriculum.

knowledge in the technical core is linked to the acquisition of identities as engineering practitioners and broader projects of service.

MULTIPLYING ENGINEERING SERVICE

A remarkable feature of these personal geographies is that they document how adding identities outside countries also led international and global engineering educators to add new broader projects—new types of service–to their pedagogy educating engineers. Daniel Hirleman and Alan Parkinson link themselves most closely to the goal of enhancing/defending American competitiveness, yet it is instructive how they write about this connection. They are not simply arguing for preparing engineers to help U.S.-based companies maximize self-interest in the expectation that good will come to all. Daniel is trying to build expansive engineers who will facilitate collaborations across the borders of countries, helping to achieve, in a way, a kind of capitalism that depends on collaboration as much as competition. Alan certainly foregrounds an image of his country at risk from foreign economic threats, especially China. Yet he also explains how his work in international and global engineering education has forced his attention to humanitarian work, leading him to reexamine his own assumptions about the broader goals of engineering service.

Other engineering participants are quite explicit about their commitments to making visible ends and projects of engineering work that go beyond serving as technical problem solvers in U.S.-based consumer industries or the U.S. defense industry. Margie Pinnell is looking for ways of promoting peace through engineering. Jim Mihelcic has dedicated himself to building more sustainable environmental practices. Anu Ramaswami seeks deeper involvement of engineers in local communities while also persuading local communities to address global problems. Rick Vaz wants to make sure engineering work goes beyond technical problem solving to incorporate attention to social, political and cultural issues for those with stakes in engineering work. Linda Phillips wants students to question what they design so they won't find themselves wondering why they are paving open spaces to build shopping centers, as she had. Les Gerhardt insists that engineering serves not only the country, so the educators of engineers must refocus their attention beyond its boundaries. Joe Mook does not explicitly articulate the broader project or projects he is trying to serve, but his personal geography strongly suggests a goal–helping U.S. engineers recognize and reflect on how much their perspectives are connected to their country so career pathways become more frequently the product of choice rather than assumption.

The non-engineers and hybrids are also explicit in seeking to help engineers examine the vision they have of engineering service by looking beyond not only consumer and defense industries but also the boundaries of the home country itself. Seeking to overcome the image of the "ugly American," Phil McKnight maintains that engineers will do work differently when they accept learning a foreign language at a high level of competence, for they will better understand themselves and achieve an enhanced appreciation for others. Bernd Widdig wants to help students use study and work outside the country to help them gain greater appreciation and tolerance for the messiness of real life, which he sees as the essence of education. Mike Nugent wants engineers and other content experts to be able to live and work amidst a multitude of differences and, perhaps most importantly, see and understand themselves as doing just that. John Grandin is firm in the conviction that by learning to unlearn many things about engineering and life they have taken for granted, they will be better able to reflect and decide on just how they want to make a difference in the world through their work. Gayle Elliott is thrilled to help immature first-year students with limited visions of themselves in the world open their eyes to the multitude of possible pathways through which they can productively make contributions through engineering work. Juan Lucena wants students to be able to recognize and analyze the political (or power) dimensions of the career trajectories they are undertaking. Gary Downey looks for ways of making it a routine practice for engineering students to examine how gaining engineering knowledge through specific courses tends to link them to some broader projects and not others.

For these sixteen educators, the experiences of adding identities outside countries set them on pathways to critically assess core practices of engineering pedagogy and work to help engineering students gain forms of knowledge and types of engineering service they would likely not otherwise recognize as even possible.

Are they tilting at windmills?

THE AMBIGUITY IN ECONOMIC COMPETITIVENESS

All these projects extend beyond the project of economic competitiveness that has provided the main force behind calls for international and global engineering education at the National Science Foundation, National Academy of Engineering, and elsewhere. To understand the relationship between the broader projects of these international and global engineering educators and the image of education for economic competitiveness, it is important to have some idea how and why the project of economic competitiveness could have become so important to engineering educators in the first place.

It has to do with the recurrent tendency of engineering educators to link their work to the progress of the country as a whole.[79] I have frequently heard (especially in conference discussions) knowledgeable observers of engineering education in the United States report that education in engineering undergoes reform efforts more frequently than in any other arena of higher education. The impression one might gain from this insight is that engineering education keeps giving the appearance of change but does not really change all that much at all. This impression misses an important dynamic in the reform activities. The changes that scale up to dominance in every case (including decisions not to change) are significant because they are in part the outcomes of struggles to keep engineering formation in step with the country. To see this, one has to look at the relationship between practices of engineering education and training and images of progress that scale up to dominance across the country. Every time the country changes direction in some significant way

---

[79]It may be more accurate to say engineering educators link their work to the progress of "humanity" as a whole. I refer to the progress of country more specifically since what counts as progress varies substantially from one country to another Downey and Lucena, "Knowledge and Professional Identity in Engineering," 2004a.

engineering educators get nervous about the knowledge emphasized in their curricula and get to work re-organizing it. That work is the key dynamic.

Engineering education in the United States first became stabilized in schools during the same period that a distinctive image of progress as low-cost production in the private sector for mass consumption scaled up to dominance across the country.[80] The period was the late 19th century, after the Civil War had settled the dispute over competing images of the Union.[81] The new land-grant schools rapidly became the major site for producing engineers to work in private industry. Prior to the Civil War, many approaches to engineering formation had competed unsuccessfully for dominance, including apprenticeship training, national military schools, independent polytechnics, programs within classical colleges, state-level military academies, and shop training. None of these practices fit well the dominant image of "low cost, mass use" and its associated industrial practices.

To say that engineering educators built curricula to fit low cost, mass use is not to say that the education of engineers served no other broader projects (e.g., American expansionism) or that struggles to build other projects into engineering education did not exist. Rather it is to say that the production of engineers for private industry whose goal was to achieve low-cost production for mass consumption became a given. It was taken for granted. It needed no justification or explanation. The new schools established curricula that prepared engineers to serve as employees in industry. The formation of engineers was responding to a newly-dominant "metric of progress" in the United States, one that set it off from pathways in other countries. In this sense, engineering educators were both cutting-edge leaders and followers at the same time. They were leading educational activities to follow low-cost, mass use.[82]

The country changed direction in the twentieth century. I abbreviate the story at the risk of caricature. In the early twentieth century (especially the 1920s), large industrial corporations successfully achieved broad acceptance for their self-described identity as key vehicles for American progress more generally.[83] Rejecting the efforts of some educators to establish an organized, rational country-wide structure for engineering formation, engineering schools continued to adapt their curricula to keep in step with the decisively prominent role of corporate industry. They took care, for example, to prepare engineers for employment within large bureaucracies by including education in the humanities and social sciences (in contrast with European schools).

After World War II, the image of Communism threatening the country's identity as a home for low-cost production for mass consumption became a given. Acceptance of the image made possible

---

[80]The following is an abbreviated version of the argument presented in Downey, "Low Cost, Mass Use," 2007.

[81]In this sense, the United States did not fully become a country until after the Civil War, for no single image of progress had gained sufficient acceptance to become dominant.

[82]For examples of accounts that draw on this argument to describe the early emergence of engineering education in Mexico, Colombia, and Brazil respectively, see Lucena, "Imagining Nation, Envisioning Progress: Emperor, Agricultural Elites, and Imperial Ministers in Search of Engineers in 19th Century Brazil," 2009; Lucena, "*De Criollos a Mexicanos*: Engineers' Identity and the Construction of Mexico," 2007; Valderrama et al., "Engineers' Identity and Engineering Education in Colombia, 1887-1972," 2009.

[83]Just as Ford's Model T dramatically expanded access to automotive transportation, General Motors later succeeded in linking different types of vehicles to distinct market segments. General Electric successfully argued that progress was its most important business, and by the 1940s few questions that what was good for General Motors was good for the country.

and justified (among many things) the work of building what became known as the "defense industry," i.e., an entirely new set of industrial practices that lay officially in the private sector but was publicly funded and aimed at confronting the threat to the country's dominant image of progress. During the 1950s and 1960s, engineering educators linked their work to these practices by re-building the technical core of engineering curricula around the engineering sciences. The dominant image of the Communist threat characterized it as a force built on Soviet science.[84] Engineering educators thus jumped onto the crowded bandwagon of science. The key point is that by preparing engineers to serve both in industries producing goods for mass consumption and in the defense industries that helped protect them, engineering educators could be confident of a tight linkage between the formation of engineers and the broader project of serving the country. They served the country through its private, but country-based, industries.

The decline of the Cold War and rise of the image of economic competitiveness as a measure of welfare and progress for the country and its inhabitants introduced a fundamental ambiguity into this broader project of service to the country. In their never-ending search for lower costs (especially via lower labor costs), industrial corporations based in the United States began to spread. They went offshore. They built new manufacturing divisions in other countries. They went international, especially after 2000.[85]

For engineering educators, the internationalization of engineering work, or flow of engineers beyond the boundary of the country became a qualitatively different challenge than constantly repositioning engineers *within* the country to meet changing demands and evolving metrics of progress. The internationalization of engineering work broke the link that engineering educators have long tended to assume between engineers working for their employers, their country, and humanity as a whole.[86]

Pursuing career pathways that carry engineers beyond the boundaries of the country makes it more difficult to determine which country's ends are served by a particular pathway. If, for example, I am an American engineer working for the Japanese company YKK in the United States, am I working on behalf of Japan or the company's host country, the United States? Or both? Or neither? If I am an American engineer working for Siemens or Michelin, does that mean I am in some sense becoming German or French? Does becoming proficient in international industrial contexts, whether outside or inside the United States, mean I have to abandon or consider less important the identity that links me to my country? Do I in fact still want my engineering work to be a vehicle for contributing to my country?[87]

---

[84] Lucena, *Defending the Nation*, 2005.

[85] Speaking at the 8th Annual International Colloquium on International Engineering Education, the President and CEO of YKK Corporation of America, the Japanese-owned manufacturer of zippers, described the departure from the United States of manufacturing plants for Levi and Lee blue jeans. Between 1999 and 2005, the number of Levi plants in the United States declined from 33 to 1 and the number of Lee plants declined from 32 to 0. He was speaking at the 2005 Colloquium for Global Engineering Education in Atlanta, Georgia (personal observation).

[86] Downey and Lucena, "National Identities in Multinational Worlds: Engineers and 'Engineering Cultures,'" 2004a.

[87] These sorts of questions are relatively new only for engineers from the industrialized countries of Europe, North America, and Asia. For several decades, students and engineers from poorer countries have traveled to the industrialized countries for higher education in engineering, after which they have had to decide whether to stay, return home, or move to yet another location.

The internationalization of engineering work has introduced a crack in the connection between the education of engineers and private sector, low-cost production for mass consumption. This connection is no longer an absolute given.[88] The change makes its more reasonable to ask "What is international and global engineering education for?" without assuming it's about advancing or defending low cost, mass use in the United States.

The sixteen contributors to this volume pry open this crack by when they challenge students to confront their own knowledge and commitments as prospective engineers. As the chapter by Jesiek and Beddoes shows, the learning experiences they offer through international and global engineering education are for the most part not new. Their practices are motivated by their own life experiences adding identities outside home countries and grounded in sober analysis. Might the contemporary ambiguity in low cost, mass use offer them unique opportunities to push the boundaries of dominant pedagogies in engineering and multiply pathways for engineering service?[89]

THE FRAGILITY OF WORK ON THE PERIPHERY

The magnitude of the challenges these educators face in scaling up their curricular projects is illustrated graphically by the fact that, at this writing, their practices of international and global engineering education have been confined to the periphery of engineering curricula (or to practices that are difficult to scale up, such as the unique curricular practices at Worcester Polytechnic Institute or small four-year trajectories that travel alongside existing curricula). Consider some typical examples, reported in general terms.

International enrollments through study abroad and academic exchange tend to be co-curricular in the sense that engineering students venture outside their curricula to experience them. These adventurous students must then accept the burden of seeking credit transfers to integrate these experiences back intro their curricula and, hence, onto their degree transcripts. Credit transfers become a key site of incongruence between the curriculum and the transcript.

International internships and co-op experiences tend to be supplemental to engineering curricula. Most occur during summer months when schools are out of session. At institutions offering co-operative (co-op) education, these experiences fit by occupying planned gaps in at-home class-

---

For example, an Egyptian engineer who received his Ph.D. in electrical engineering from the University of California at Davis reported in an interview (confidential interview, Cairo, July 2003) that he never bought a house during his eight-year stay in the United States to remind himself that his plan was to return to Egypt and help his country. Another Egyptian engineer found his decision to return to Egypt especially affirmed when he visited a fellow countryman living an affluent life in Silicon Valley, California (confidential interview, Cairo, July 2003). As they discussed their homeland, the man's young daughter happened to enter the room. He began to cry.

[88] One indicator of this separation and continuing ambiguity is the National Academy of Engineering's Grand Challenges project, funding by the National Science Foundation in 2006 and announced publicly in early 2008. In relation to this project, the fourteen Grand Challenges are notable because they are explicitly service projects for the world as a whole and fulfilling them necessarily re-orients engineering service beyond private employers. To "provide access to clean water" or "restore and improve urban infrastructure" is not the same, for example, as seeking economic supremacy. See www.engineeringchallenges.org.

[89] The projects described in this volume are but a small subset of what educators seeking to transform dominant practices of engineering pedagogy to scale up (an especially large arena of energetic work is design education).

room instruction. Yet, these experiences either do not appear on the transcript at all or receive mention as extras.

International projects, service learning projects abroad, and collaborative research projects with partners abroad tend to take place either during school holidays or as part of senior capstone design experiences. International projects and service learning projects abroad are generally supplemental to engineering curricula, but students may receive course credit. Capstone design experiences are already fully integrated into engineering curricula, but at the margins. They tend to sit on the boundary between the engineering science core and graduation. As a result, they tend to be more flexible and can be adapted by interested faculty for interested students.

Double major or dual-degree programs gain prominent positions on students' transcripts as well as on diplomas. The international experience, e.g., a degree in a foreign language and literature, tends to fall, however, outside the engineering portion of the transcript. Also, administrators of the second major or degree typically loosen its requirements to make it more attractive to engineering students (e.g., by double counting some course for both majors or degrees).

Certificates and minors constitute significant interventions in the existing curriculum, affording students an opportunity to adapt both technical and nontechnical electives toward international experiences and knowledge. The cost to students is they must plan their transcripts at an early point in order to gain entry to the courses that afford coherent learning experiences. Still, because certificates and minors are defined in relation to a core or major, they invite students to identify relationships between them.

Elective courses supporting the international education of engineers fall on the peripheries of the existing engineering curriculum and show up as line items on the transcript. Their long-term value to students depends upon the will and capacities of students to integrate practices from such courses into their professional practices.

As exceptionally energetic and creative agents working the periphery, the majority of contributors to this volume describe remarkably solitary work building small-scale networks of people and institutions through extended personal contacts.[90] The work of building and maintaining new networks typically requires far more time than agents of change expect or may be authorized by released time or salary support. The time committed to establishing a new program in international and global engineering education is thus often time away from other responsibilities.

Building and maintaining a program locally requires new allocations of resources. It requires physical space and access to staff support. It requires advertising to make the initiative known to students. It requires means for achieving direct contacts with students to help them figure out how to fit international and global engineering education into their programs of study.

Network building requires establishing and maintaining personal relations with a variety of colleagues and officials based in other countries. The work is often painstaking and ranges from extensive telephone and email communications to personally hosting visitors and frequent travel

---

[90] Even Alan Parkinson, working as engineering dean, reports difficulties in making international and global engineering education a focus of his faculty.

for network maintenance. The reliance of these networks upon personal relations both near and far makes activities in international and global engineering education particularly vulnerable to changes in personnel. The simple replacement of one person can bring down an entire enterprise.

One implication of the relative marginality of existing programs in international engineering education to the curricular core is that their elimination would pose little hazard to continued delivery of the curriculum. Another is that relatively few students take advantage of them. Existing international programs typically reach no more than 5% of undergraduate engineering students, and often fewer than 2%. The first problem of scale in international and global engineering education is the survival of existing practices.

ENGINEERING KNOWLEDGE BEYOND PROBLEM SOLVING?

International and global engineering educators face an especially difficult knowledge problem. The separation between core and periphery is the distinction that separates the knowledge and practices of engineering students from those of other students. The specific practices of problem solving also separate students from one another in different engineering majors (e.g., mechanical, civil, chemical, electrical, etc.). Working on the periphery, international and global engineering educators face two related challenges. One is to specify the knowledge contents of the learning experiences they seek to share with students. This is why Kacey Beddoes and I asked participants to describe in their revisions the knowledge contents of their experiences: What did they know and what could they do after key experiences of transformation that they didn't know or couldn't do beforehand? This is no easy question to answer, and we all struggled. The second challenge is to persuade engineering science faculty and other engineers that this new knowledge and these new practices could be (or are in fact) *engineering* knowledge and practices.

International and global engineering educators unwittingly exacerbate the problem of knowledge in international and global engineering education by characterizing the new knowledge as cultural in content. Personal geographies in this volume regularly mobilize the term "culture" to refer generally to forms of life in other countries. Culture is linked to language, literature, society, history. The practice of adding identities outside countries is often characterized as engaging people from other cultures. Education becomes a problem of intercultural sensitivity and competence. Gaining such sensitivity and competence is crucially important, practitioners insist.

This use of the term culture is worrisome. International and global engineering educators are actually joining colleagues in the engineering sciences when they portray cultural knowledge as supplemental, external to engineering. The anthropologist Sharon Traweek once characterized the discipline and community of physics as a "culture of no culture."[91] The same might be said for engineering. Practitioners of the technical core in engineering fields, including its budding apprentices, understand its contents in part by the absence of culture. Culture is stuff to be added later.

[91]Traweek, *Beamtimes and Lifetimes: The World of High Energy Physicists*, 1988. See also Gershon and Taylor, "Introduction to 'in Focus: Culture in the Spaces of No Culture'," 2008.

In recent years, I have been asking engineering faculty how easy or difficult it is to characterize engineers as members of different cultures. They regularly assert it is increasingly difficult. Characterizing a person as a member of a culture typically depends on the assumption that cultures are membership groups that are discrete, distinct from one another, and have boundaries that overlap roughly with the boundaries of countries. Thus, someone who grew up in a given country presumably is a member of that country's culture and thus has a cultural identity defined more or less in national terms. But this assumption is challenged by the rapidly increasing mobility of populations across national borders and high level of diversity within them. Such mobility means that people increasingly have added identities that locate or root them in more than one country. Individual cases become quite complicated, especially as people spend substantial periods of their lives in countries outside the country of birth. At one prominent engineering research university, for example, I was told more than 50% of engineering graduate students and one-third of engineering faculty were born in other countries.[92]

As a result, the idea of cultures as membership groups overlapping with countries, a holdover from the mid-20th century, becomes too simplistic to characterize differences among people in the present. If people cannot easily be described as members of single cultures, then equating cultures with countries and classifying residents in a given country as members of its culture further undermines the effectiveness of emphasizing cultural knowledge in international and global engineering education. The emphasis on cultural knowledge is a self-limiting distraction in international and global engineering education when it is defined as external to engineering knowledge.[93]

What these personal geographies do show is that a key step in learning to reflect critically on engineering education is adding identities outside the *country*. Because engineering educators have linked dominant practices of engineering education to the trajectories of countries, it is leaving the country that crucially challenges engineers and engineering students to reflect on and rethink their knowledge and commitments.

One can see, upon closer inspection, that statements about the benefits of global learning for engineering students typically locate those benefits in encountering and coming to understand engineers and other potential co-workers who are raised, educated, and living in countries other than their own. The innumerable calls over the past decade for global learning for engineers and the wealth of emerging initiatives in international education tend to make encountering people who are raised, educated, and living in different countries, especially engineers, a key target group. Although such people comprise only one subset of the configurations of identities and experiences that engineering students are likely to encounter on the job, their special educational status is an indicator of the goal international and global engineering educators tend to advance: learning to engage effectively ways of understanding and practicing engineering work that differ from your own.

Yet when leaving the country is not an easy option for students, how are educators in international and global engineering education to make any headway in scaling up critical awareness of

[92]Confidential conversation.

[93]Developing the course Engineering Cultures (see my personal geography) was an effort to resist locating culture outside engineering knowledge.

dominant practices in engineering pedagogy? Perhaps it may be helpful to highlight that existing dominant practice instill knowledge and commitments at the same time.

ADDING IDENTITIES TO ENGINEERING STUDENTS: A TRANSCRIPT VIEW

The integrated degree path view of the curriculum described above is a faculty view. Student experiences of the curriculum, by contrast, tend to be built around check sheets and transcripts (see Figure 18.2). Students ask: What do I need to complete in order to gain the credential of the degree? Have I taken the right prerequisites? Which technical electives should I take? Which free electives should I take? How do I work in a co-op job? How do I transfer in credits from study abroad? Am I on track? Will I graduate on time?

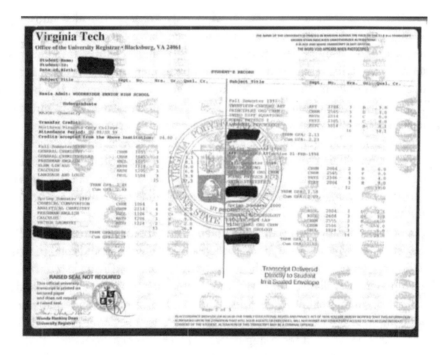

Figure 18.2: A Transcript (Student) View of the Engineering Curriculum.

In piecing together transcripts, students are adding both knowledge and commitments to broader social projects. One way to see this is to look at the significance attributed to technical electives as vehicles of specialization. Supporting material for the degree path described above tells students, "Technical electives should be selected to match your career interests." "Popular areas," it continues, are "automotive engineering, biomechanics/biomedical engineering, green engineering, industrial design, industrial and systems engineering, machine design, materials science, mechanics, nuclear engineering, robotics, thermal/fluids engineering." It then calls particular attention to the fact students will soon be looking for jobs, invoking the identities of employers and picturing career trajectories into them:

> Note that prospective employers will be more impressed if you have taken courses that meet their needs than if they saw an unrelated minor on your resume. You can always bring attention to your choice of technical electives on your resume by including a statement such as: 'I concentrated in automotive engineering by taking the following courses . . . .' (That would certainly get an auto manufacturer's attention more than a math minor would!). If you are not sure what area you would like to specialize in, you should consider taking general technical electives in a variety of areas and pursue engineering internships or co-op jobs as a means of exploring potential careers.

In this image, technical electives initiate the process of adding specific industrial identities to engineering students.[94] The student who specializes in automotive engineering is already adding an identity linked to the auto industry. While taking this step does not prevent the addition of different identities later, it does make any later redirections away from that industry a definite change that might require new training. Technical electives matter in the development of practitioner identities.

Senior capstone design courses also add identities to students by integrating them into specific broader projects of engineering service. At this writing, students pursuing the mechanical engineering degree could choose from among nearly fifty projects offered by their faculty. Every project includes a maximum number of participating students, and levels of participation offer one indicator of the desirability of specific identities. Consider these in descending order of permitted participation:

> hybrid electric vehicle (27); Baja off-road vehicle; biomedical engineering (25); Formula race car; MANTis on-road/off-road vehicle; pediatric medical devices (20); new vehicle for Mexican market (16); aerial surveillance vehicle; electric vehicle wind turbine (15); jet propulsion; drag in turbulent flow (12); facial robotics (11); human-powered vehicle; turbocharger test stand; nuclear engineering; small-scale windmill; tobacco harvesting; unmanned ground vehicle; bio-inspired sonar; autonomous surface vehicle competition; sea water electrolysis; solar power for remote medical clinics; solar energy; wind power (10); off-road propulsion; artificial hand; anti-submarine vehicles (9); ultrasound imaging vest; steerable audio speaker; blind driver; humanoid robot; hybrid vehicle motor (8); piezoelectric testing machine; solar house (7); tobacco curing; sporting helmets; aerial camera; aircraft tracking; pursuit/evasion tactics; global emerging markets automobile; Rolls Royce project (6); robotics for high-school students; desktop fabrication;

---

[94]Note that second and third-year language courses in French, German, Russian, and Spanish all count as technical electives "if 9 additional engineering credits (taught in [language]) are earned at a foreign educational institution after completion of these language courses."

cam-based transmission; autonomous supply vehicle; modeling football defenses (5); remote-controlled aircraft; traffic injury prevention (4); biodiesel fuel; off-road lifts (not indicated).

Senior design projects matter in the addition of new identities as engineering practitioners.

Core engineering science courses also matter in the addition of practitioner identities, even if more subtly so. Engineering science courses introduce students to abstract mathematical arenas that only partly overlap with one another. Engineering statics, dynamics, kinematics, and deformable bodies all introduce students to what the MIT engineer Louis Bucciarelli has called "object worlds."[95] Bucciarelli's point is that each engineering science creates and lies in one or more object worlds that engineers must enter into (and add to their identities) to do their analyses. The mathematical objects in these worlds are both crucial to quality engineering work and a significant source of difference and disagreement among engineers.[96] "In the simplest terms," Bucciarelli writes, "design is the intersection of object worlds." Systematically examining three design projects that experienced high levels of uncertainty, Bucciarelli finds that "[t]he apparent incoherence and uncertainty of the process[es] . . . derives in large measure from the differing interests and viewpoints of different parties to the design." He observes how engineers and other professionals working within different object worlds "will construct different stories according to their responsibilities and . . . technical, professional interests." As a result, because "the authors of these stories display full confidence in their construction," the key issue in defining the engineering problem at stake is not overcoming uncertainty but reconciling different perspectives.[97]

Dominant pedagogical strategies in each engineering science, consisting of lectures, problem sets, and exams, tend to focus on making sure students demonstrate an ability to implement its practices. Less often do they challenge students to articulate the value of those practices or explain how they are distinct from other practices. Rarely, for example, does a thermodynamics class in mechanical engineering address how it differs from the thermodynamics taught in chemical engineering. It would be highly unusual to ask students such questions as: What are the key entities and processes in this thermodynamics course and how do they relate to one other? How are these entities and processes similar to or different from those in the heat transfer course students take? How do thermodynamics and heat transfer connect to one another, or not? What objects does one see and practices does one gain by invoking first-law or second-law equations? Yet it is in beginning to ask such questions, and others like them, that practices of critical self-analysis move to the heart of engineering pedagogy and, hence, engineering knowledge and practice.

In completing engineering science courses students are challenged to integrate the diverse objects they encounter into their own identities. Years ago, Juan Lucena and I joked how learning engineering statics led us to begin analyzing everything in our lives in terms of forces, including relationships with other people. We were adjusting to the incongruence between a newly-added

[95] Bucciarelli, *Designing Engineers*, 1994.
[96] Engineering science faculty commit their careers to advancing and improving the abstractions that constitute these arenas. In the process, they add identities linked to the promise and value of these theoretical objects.
[97] Bucciarelli, *Designing Engineers*, 1994, 20, 51, 71, 72.

identity as an engineering practitioner and existing personal identities. I myself still carry a nagging feeling that my practitioner identity as a mechanical engineer is incomplete because I persuaded my department to allow me to substitute courses in environmental engineering for that required heat transfer course. Asking questions about whom one by means of the engineering sciences brings into the technical core of engineering curricula practices that now reside largely on the periphery.

By challenging students to ask new questions about their knowledge and commitments, the international and global engineering educators in this volume are, in effect, viewing the formation of engineers through learner-oriented perspectives. Further scaling up transcript views of student learning would call attention to just how different engineers are from one another, even within a given field. Must a degreed mechanical engineer (or electrical engineer, chemical engineer, etc.) be seen as just one thing? It is likely no two students have exactly the same transcript, with all the same courses and all the same instructors. It is certainly the case no two students add the same configurations of professional and personal identities as they move through engineering curricula, especially when one begins to take co-curricular experiences into consideration. From a transcript viewpoint, engineering students are acquiring knowledge, practices, and identities as they move through engineering curricula, adding them to distinct knowledge, practices, and identities they brought with them into those curricula. In a sense, a transcript view makes every instructor who shows up on the transcript of an engineer effectively an engineering educator.

Viewing engineering education up through the student transcript, as the addition of multiple identities, rather than down through the degree path sheet, as the acquisition of a single identity, redraws the landscape of possible futures for engineering curricula. The top-down view of a packed curriculum frequently produces the knee-jerk response to any efforts to add new knowledge, practices, and identities, "What do you want to cut?," while leaving intact the arrangement of core and peripheral experiences. A bottom-up view through the transcript, by contrast, challenges educators to justify anew every required and elective component for each and every student, questioning its givenness. Why is that course there? What makes the knowledge, practices, and identities it adds through its new objects so crucial or important? What other configurations of theoretical objects, practices, and identities might be justifiable?

As just one thought experiment, imagine the dominant practices in the core as constituting but a single track in a departmental degree program that includes other, new tracks. That is, the curriculum that is placing highest emphasis on core learning in the engineering sciences becomes an engineering science track, structured perhaps especially to prepare students for research positions or graduate school. Consider again the three strands of mechanics, electrical/instrumentation, and thermal/fluid sciences on the degree path sheet. Given the many design projects in biomedical engineering, would it make sense to develop an additional strand involving the biological sciences? Would graduating students still be mechanical engineers if they selected three out of four strands of engineering sciences rather than taking the existing three strands for granted and supplementing them with other knowledge, practices, and identities?

In like fashion, an engineering design track could include coursework in industrial design, architecture, or other design disciplines, preparing students for careers emphasizing design work. An engineering and management track would specifically help students prepare for the work of collaborative problem definition in private industry, especially by training them to analyze the types of knowledge other non-engineering managers possess and use. An engineering and policy track or engineering and society track would prepare students for the work of collaborative problem definition beyond the firm, e.g., in government or non-profit sectors. Extrapolating the idea, a multi-field general engineering track, degree, or possibly advanced degree program could introduce students sufficiently to a range of fields to enable them to function effectively as mediators among different types of engineering specialists.

A significant challenge in using a bottom-up, transcript viewpoint to imagine alternative curricular practices is that it renders more porous the boundaries around engineering fields. When is a mechanical engineer still a mechanical engineer? This is a significant question. But it is also not a novel question. The boundaries of engineering fields are already becoming more porous as they link themselves to emerging biotechnologies, information technologies, and nanotechnologies.[98]

A significant benefit in multiplying tracks could be to make every course a site for practicing critical analysis of one's own knowledge and commitments. It becomes possible to picture engineering science faculty having to compete for students rather than simply assuming they will be there. The work of effective pedagogy could begin to include sharing both why a particular engineering science is important and how it is distinct from other engineering sciences. Since every curriculum would be a track, it would have to acknowledge more explicitly what it leaves out, which in turn could help integrate more into the visions and planning of students images of continuing their education post-graduation. And for the learning projects in this collection, neither engineering faculty nor engineering students could take it for granted that the benefits of international and global engineering education fall on the periphery rather than in the core.[99]

The possibility of scaling up shifts in dominant practices of engineering science pedagogy increased with the year 2000 switch in accreditation practices from counting credits to measuring capabilities. Yet adjustments in accreditation practices by no means compel changes in everyday teaching practices. Indeed, the very expectation of accreditation can inhibit even the most adventurous efforts to add new identities to engineers through innovative pedagogy.[100]

[98]Williams, *Retooling*, 2002.
[99]I have elsewhere built on a transcript view to argue for expanding the core outcomes of engineering pedagogy from "problem solving" to "problem definition and solution." See Downey, "Keynote Address: Are Engineers Losing Control of Technology?," 2005, The point is that collaborative problem definition with both non-engineers and engineers is an integral part of engineering work left out of engineering pedagogy, including work on "problem formulation" in design courses. Collaborative problem definition depends upon expecting other stakeholders in engineering work to have both knowledge and value, and its practices rely upon questioning location, knowledge, and desire.
[100]Seron and Silbey, "The Dialectic between Expertise Knowledge and Professional Discretion: Accreditation, Social Control, and the Limits of Instrumental Logic," 2009.

REDEFINING THE CORE

The image of the globally competent engineer that appears to be scaling up to dominance, both in the United States and beyond, is linked tightly to the associated broader project of service through employment in multinational industries. For U.S. engineers in particular, adding this career identity may seem like a relatively straightforward extrapolation of service to low-cost mass use across the country. Yet what the personal geographies in this volume clearly demonstrate is that linking global competence to service through multinational industries not only construes far too narrowly the new forms of knowledge that come with adding identities outside home countries but also fails to capture the importance of confronting discrepant moments that challenge established identities.

In adding identities outside countries, contributors to this volume began to see themselves differently. They learned to recognize that what they had always taken for granted in both personal and professional practices could have been different, and indeed became different in the process. The big payoff was in learning to analyze their own locations, forms of knowledge, and desires in relation to those of others. Such learning introduced, in turn, the very possibility of scaling up other practices in engineering pedagogy in addition to, alongside, or in place of those that have been dominant.

These educators must confront barriers posed by the given separation of technical core and nontechnical periphery. It seems unlikely international and global engineering education will be sufficient to build into the core of engineering pedagogy discussion and debate over what sorts of knowledge, identities, and broader forms of service each course or learning experience facilitates or not. But will such practices ever become routine without it?

REFERENCES

Bucciarelli, Louis L. *Designing Engineers.* Cambridge, Mass.: MIT Press, 1994. 428

Downey, Gary Lee. *The Machine in Me: An Anthropologist Sits among Computer Engineers.* New York: Routledge, 1998. 416

Downey, Gary Lee. "Keynote Address: Are Engineers Losing Control of Technology?: From "Problem Solving" to "Problem Definition and Solution" in Engineering Education." *Chemical Engineering Research and Design* 83, no. A8 (2005): 1–12. 430

Downey, Gary Lee. "Low Cost, Mass Use: American Engineers and the Metrics of Progress." *History and Technology* 22, no. 3 (2007): 289–308. DOI: 10.1080/07341510701300387 420

Downey, Gary Lee and Juan Lucena. "Knowledge and Professional Identity in Engineering: Code-Switching and the Metrics of Progress." *History and Technology* 20, no. 4 (2004): 393–420. DOI: 10.1080/0734151042000304358 419, 421

Downey, Gary Lee and Juan C. Lucena. "Engineering Selves: Hiring in to a Contested Field of Education." In *Cyborgs and Citadels: Anthropological Interventions in Emerging Sciences and Tech-*

nologies, edited by Downey, Gary Lee and Joseph Dumit, 117–142. Santa Fe: School of American Research Press, 1997. 416

Downey, Gary Lee and Juan C. Lucena. "National Identities in Multinational Worlds: Engineers and 'Engineering Cultures." Paper presented at the 9th annual World Conference on Continuing Education for Engineers, Tokyo, Japan, 2004. DOI: 10.1504/IJCEELL.2005.007714

Gershon, Ilana and Janelle S. Taylor. "Introduction to 'in Focus: Culture in the Spaces of No Culture'." *American Anthropologist* 110, no. 4 (2008): 417–421. DOI: 10.1111/j.1548-1433.2008.00074.x 424

Lucena, Juan. *Defending the Nation: U.S. Policy making to Create Scientists and Engineers from Sputnik to the 'War against Terrorism'.* Lanham, Maryland: University Press of America, Inc., 2005. 421

Lucena, Juan. "Imagining Nation, Envisioning Progress: Emperor, Agricultural Elites, and Imperial Ministers in Search of Engineers in 19th Century Brazil." *Engineering Studies* 1, no. 3 (2009): DOI: 10.1080/19378620903225067 420

Lucena, Juan C. "*De Criollos a Mexicanos*: Engineers' Identity and the Construction of Mexico." *History and Technology* 23, no. 3 (2007): 275 - 288. DOI: 10.1080/07341510701300361 420

Seron, Carroll and Susan Silbey. "The Dialectic between Expertise Knowledge and Professional Discretion: Accreditation, Social Control, and the Limits of Instrumental Logic." *Engineering Studies* 1, no. 2 (2009): DOI: 10.1080/19378620902902351 430

Traweek, Sharon. *Beamtimes and Lifetimes: The World of High Energy Physicists.* Cambridge, Massachusetts: Harvard University Press, 1988. 424

Valderrama, Andrés, Idelman Mejía, Antonio Mejía, Ernesto Lleras, Antonio Garcia and Juan Camargo. "Engineers' Identity and Engineering Education in Colombia, 1887–1972." *Technology and Culture*, no. (2009): DOI: 10.1353/tech.0.0341 420

Williams, Rosalind H. *Retooling: A Historian Confronts Technological Change.* Cambridge, Mass.: MIT Press, 2002. 430

APPENDIX A

Making Explicit Diverse Trajectories

Gary Lee Downey

Near the end of the workshop, a senior engineer participant approached me. "I have to tell you." he said, "Before the workshop, I was wondering 'What's all this one-minute-here, two-minutes-there bullshit.' But it really works! I now know things about myself I never thought about before!"

Dr. Pimmel's vision was of a process leading to a persuasive report. Judging from the multitude of reports on engineering education that have appeared since the mid-1980s, it is perhaps fair to say that the typical study practice involves bringing together demonstrated leaders in a given area to define a problem or ambiguity in the present, negotiate an image of a desirable future (e.g., the "engineer of 2020"), and specify courses of action to achieve that state, all in a jointly-authored document. The process mobilizes both authority and analysis. The hope is that the force of their united stance and considered judgment will persuade readers to accept the accuracy of the problem definition, attractiveness of the defined future, and appropriateness of the recommended practices for achieving it. The report is about leadership.

Yet leadership is an outcome dependent on followers, and all too many reports have short half-lives. The whole process of negotiating definitive findings and recommendations involves eliding or ignoring differences in position and perspective through practices designed to produce agreements. The expectation is that a common diagnosis of the present can be produced, and by agreeing to serve participants are also basically agreeing to agree. By aiming at single, collective authorship, a consensus report on international and global engineering education could risk actively hiding differences that could prove crucial in how readers respond to it.[1] The report could become a megaphone for a single stance or limited set of stances and unwittingly generate resistance or opposition from those who stand elsewhere rather than win their interest and acceptance. Participants and readers who find their identities and perspectives represented incompletely in or in conflict with the grounding diagnosis may simply ignore the findings and recommendations. If they don't agree on what's at stake in the present, it's not likely they'll subscribe to an image of a particular future nor align themselves with the recommended pathways for heading there. They continue going about their business, and the

---

[1] See the Jesiek and Beddoes background chapter for an account of how the report of the 2008 Summit on Global Engineering Education works to preserve the force of consensus amidst diverse trajectories by advancing multiple objectives in parallel with one another.

force of the report recedes quickly into the past. The decline of a consensus report was the first thing I pictured in discussions with Dr. Pimmel about a workshop on global engineering education.

I also pictured an alternative process that began by highlighting differences. Foregrounding differences is essential to forestalling narrowness. Also, making visible and then juxtaposing distinct commitments to broader, external projects might call attention to areas of overlap and facilitate collaborations that began with the acknowledgement of differences. Rather than planning strategies for the future, I focused on figuring out how to map contrasts in the present in international and global engineering education.

I invited Brent Jesiek (Virginia Tech/Purdue University) and Juan Lucena (Colorado School of Mines) to serve as co-organizers and co-PIs on the grant. Brent accompanied me on a trip to the 10th International Engineering Education Annual Colloquium at Purdue University to identify possible participants. He also agreed to draft a background history of international engineering education in the United States and organize the workshop evaluation. Juan researched and suggested additional names that were unfamiliar to me. He also suggested the term "personal geography" after hearing it used productively at a conference. The three of us held occasional teleconferences to discuss workshop plans, and they contributed reactions to the abstracts and reviews of most of the manuscripts. After Brent invited Kacey to co-author the background history and I saw the high quality of her work, I invited her to join the organizers and serve as co-editor of this volume. Committed to other projects, Brent and Juan elected not to serve as co-editors.

Every potential participant I telephoned readily accepted the invitation. The written letter of confirmation characterized participants as "past and present activists" in "global education for engineers."[2] Perhaps people were attracted by the intrinsic value of the project. Perhaps it was the opportunity to have their work gain new visibility through publication in a book. Maybe it helped that the workshop would take place at the main National Academies building on Constitution Avenue in Washington, D.C. The official host for the meeting, the Center for the Advancement of Scholarship on Engineering Education (CASEE) at the National Academy of Engineering, has its offices in that building.[3] They all immediately accepted the challenge to produce a personal geography despite hearing only a few clues as to what that might involve.

The formal, follow-up invitation to participate in this project invited participants to "both recount the emergence of your own perspective on global education for engineers and locate that perspective in relation to those around you." The first step in the writing process was a 500-word abstract. Brent and Juan offered comments on each one, after which I drafted a detailed 500-1,000 word response to each author. It immediately became clear that steering people toward the personal geography would be a challenge. I had mainly (and wrongly) been worried that educators accustomed to writing program documents and telling program histories would overcompensate by writing personal accounts that departed too far from the trajectories of careers. The biggest challenge for most participants (including me) was to abandon the genre of program history.

---

[2] See Appendix B for copies of all formal correspondence with the contributors.
[3] I was a Boeing Foundation fellow at CASEE from 2005 to 2007 and still serve as affiliated faculty.

Most of the abstracts resembled proposals for education conferences, generating three types of editorial responses. The first was a discomforting request for earlier moments, of the sort: How did you get there in the first place? The accounts had to include sufficient moments or episodes for readers to locate patterns in interactions among personal and work identities, recognize the significance of discrepant events, and understand the broader projects underway. The person had to become visible far earlier and to a far greater extent than most writers (including both engineers and non-engineers) were accustomed to sharing.

Gaps in a trajectory would also raise questions for readers. Missing phases in a career or missing decisions to change direction would undermine the reader's ability to assess what was considered and what was not. "Please slow down here," said one reply. "How does an engineer make this transition from signal and image processing to internationalizing engineering education?" "At what point did you gain faculty status?" asked another. "What led to the transition and what did it mean for you?"

To clarify changing positions, it was important for the personal geographies to make visible the identities and perspectives of others encountered along the way, especially those present at key moments. This was especially difficult for many participants, for the genre was asking them to become ethnographers of their own experiences. It was asking them to accept the discipline of outside observers watching them work and live. "What sorts of different perspectives were engaging one another here?" said one response. "Might you consider interviewing others about their perspectives in order to better understand and present them? Were there disagreements? What made "global awareness" significant, especially given that the other areas are likely easier to identify and calibrate?"

The formal guidelines for writing these personal geographies sought to anticipate difficulties in describing one's own career history as movement across a territory. The guidelines included a detailed introduction to the practices of questioning location, knowledge, and desire, built around an encounter between conflicting perspectives that might feel familiar. The encounter was between a "young, female advocate of study abroad for engineers" and a "skeptical older male faculty member who teaches in the engineering sciences." While the study-abroad advocate stood on the margins of the curriculum, the engineering science faculty member stood at the core. While it was difficult for her to demonstrate what knowledge was to be gained from study abroad, it was not difficult for him to explain the knowledge students gained in his classroom. And while the study-abroad advocate might envision a growing population of students interested in pushing their boundaries, the engineering science faculty member could picture only loss.

The draft manuscripts were due four months later, two months prior to the workshop itself. The guidelines for reviews anticipated the unusual format of the workshop itself, which was designed to transform critics into co-authors. "Please think of this," said the review guidelines, "as a collegial review for a friend rather than a peer review for a journal." It would be difficult for reviewers to become co-authors later if they had adopted sharply critical stances in the initial reviews. Each author wrote reviews of eight to nine other manuscripts (all members of their assigned group). "A half page is sufficient," continued the guidelines, "More is better." Then the guidelines offered a half-page of specific questions ranging from "Does it describe activities that were taking place prior to this one" to "Does it make visible struggles and uncertainties?" The responses were surprising, even dramatic, as many reviewers found themselves moved to respond not only with questions and suggestions but also lengthy commentaries.

Brent, Juan, and I submitted separate commentaries, in part to multiply the number of responses each author received and in part to guide the subsequent workshop conversations by anticipating them. In retrospect, I should have anticipated that authors would be uncomfortable departing too far from the accustomed genre of the program history. Including oneself in a program history was discomforting not only by risking narcissism but also because the narcissism could be read as naïve, misguided advocacy. It quickly became clear that the central challenge of the workshop itself would be to draw out details and moments that authors considered to be awkwardly personal yet were crucial to understanding the trajectory as a whole.

My strategy was to compose an especially lengthy reply to each manuscript. It is one thing to suggest in general terms, I realized, that the author "map more completely both people's specific perspectives and what they were working to accomplish," as I asked of Mike Nugent's manuscript. It is quite another to show specifically how and where by reproducing and querying large numbers of specific passages, such as "Given the circumstances, in many ways one could understand the views of the FIPSE staff that these programs were, if not a failure, a distraction from the core mission of FIPSE." "Could he," I asked in my commentary, "elaborate more the dominant perspective?" My hope was that by systematically reproducing and questioning the text of every tantalizing but incomplete description, allusion, or observation, I might legitimize questions aimed at highlighting normative commitments that others would like to ask of one another but didn't feel authorized to initiate. I also hoped to establish a rhythm of such queries that others would follow at the workshop itself. This strategy largely worked.

The key feature of the workshop format was that authors could not participate in the discussions of their manuscripts.[4] Each of the seventeen manuscripts was allocated a 30-minute slot, strictly enforced, and discussions took place in two workgroups of eight or nine members each. "Authors are free to give a 1-minute introduction to the manuscript and where it is going," said the instructions. In many cases, reading the reviews had already persuaded authors to take their revisions in a new direction. "You then have to sit quietly and listen for the rest of the session." Authors were permitted to return for two minutes at the end.

Each session was led by the Primary Respondent. "Your job," said the instructions, "is to . . . spend about 5 minutes summarizing the manuscript and the written comments." The responsibility here was to perform the author's perspective by "[d]escrib[ing] what the author is seeking to accomplish," as well as to remind reviewers what they wrote by "provid[ing] a brief overview of

---

[4] I had first developed the workshop practices with anthropologist Joseph Dumit in a 1990s seminar at an anthropology think tank and had since 2005 been using them successfully with the Research in Engineering Studies group at Virginia Tech (a collection of student and faculty researchers interested in engineers and engineering), as well as at a 2006 international workshop for researchers interested in engineering studies. See Downey and Dumit, *Cyborgs & Citadels: Anthropological Interventions in Emerging Sciences and Technologies*, 1997. For more on the Virginia Tech Research in Engineering Studies Group, see www.res.sts.vt.edu, and for further information on the 2006 engineering studies workshop, including video introductions of participants, see www.inesweb.org.

major strengths and possible areas for further work." "Remember," the instructions cautioned, "you are engaging unfinished work as a colleague rather than drafting a review as an editor." To get a conversation going, the Primary Respondent was followed by the Secondary Respondent, who had two minutes "to address points not raised by the Primary Respondent or offer additional emphasis in one or more areas." All group members were then responsible for contributing comments and suggestions for further developing the manuscript, called on by the Primary Respondent if necessary. After group members had each contributed, other observers (including members of the other group) were given the chance.[5]

The main objective of this format is to replace the typical author-meets-critics exchange with a discussion in which respondents necessarily assume both identities. When respondents are inclined to offer criticisms, they quickly realize that authors cannot reply. This compels the would-be critic to explain how the author might respond to the criticism. Not only are the authors silent but they are also made to disappear. The Primary Respondent is responsible for referring to the author in the third person (she or he) rather than the second (you). Although difficult to enforce in practice, when this works well the respondents collectively discuss the author in the third person. They slowly, but definitely, begin to shift identities from critic to co-author. The author, in effect, has not been silenced at all but multiplied, in this case by a factor of eight or nine. Also, and perhaps most importantly, by actively taking on and advancing the perspectives of other participants, they are teaching themselves to understand and assess their own practices and commitments from other points of view. They rethink the positions of companions in their own travels (see Appendix B for a sampling of the conversations that took place without authors participating).

Over the course of several discussions, the two groups at the workshop each established a strong sense of solidarity *qua* group members. That is, the many boundaries separating their positions and perspectives began to lose relevance in the workshop space. All participants held equal status at the table. All were Primary Respondents, Secondary Respondents, and authors. All were required both to talk and to listen. Each group discussed two manuscripts and then took a break for two manuscripts, giving them time to continue discussions and initiate new ones in the break room.[6]

Following the workshop, the authors had three months to produce an initial revision.[7] The guidelines for revisions first offered general observations to remind authors of the purpose of the project and its current state. The main goal of the volume would be to present "research results," with a focus on "the pathways traveled by the agents of change." To that end, many manuscripts would have to "be more personal," accept "the challenge of filling holes," and "describe what we learned about ourselves at key moments."

The guidelines also asked authors to reflect especially on knowledge without losing the emotion: "What did you know after a given experience that you didn't know before?" Many drafts had

---

[5] In addition to the authors of personal geographies and Brent Jesiek, other active participants included Russell Pimmel and Sheryl Sorby of the National Science Foundation.

[6] See Supplement 2 for Brent Jesiek's summary of workshop evaluations.

[7] One of the seventeen workshop participants elected not to submit a revision. Also, Brent Jesiek participated in workshop discussions, and one author was unable to attend.

called attention to location and desire but not knowledge. Contributions to the volume would have to address the question of knowledge if they were attempting to persuade others to adopt their educational practices. One reason international and global engineering education has lived in the margins is because other engineering educators and all too many engineering students have not judged its practices to be sources of essential knowledge.

The review process was completed when Kacey and I closely read, commented on, and significantly copy-edited every revision. In some cases, authors still omitted seemingly crucial career moments in which they made important changes of direction or otherwise left gaping holes. With directed, in-text comments and a summary statement, we cajoled authors one more time to accept the challenge of the personal geography. The second revision is what you have before you.[8]

---

[8]Two manuscripts went through one additional revision, at the request of the authors.

APPENDIX B

Conversations without Authors

Following is a sampling of the conversations that took place without the participation of authors. As Appendix A explains, this was the key feature of the Workshop.

The purpose of this method is threefold: (1) to replace the typical author-meets-critics exchange by asking respondents to accept the responsibilities of authorship, (2) enable participants to teach themselves to understand and assess their own practices and commitments from other points of view, and (3) build solidarity and trust among participants without erasing differences among them or otherwise forcing agreement.

The final personal geographies respond to these conversations. Including a sampling enables you to get a feel for the challenges each author faced. To facilitate each searching, they are in alphabetical order by author's last name.

CONVERSATION WITHOUT GARY DOWNEY

The author's description of his personal evolution from honors engineering student to cultural anthropologist to – well, to whatever the appropriate descriptor would be today, and I say that in awe, not in mockery – is quite interesting.

Gary asked two basic questions: What does it mean to be an engineer and what counts as engineering knowledge?

The overarching point of the paper seems to be about how to bring about change in the normative educational environment of engineering.

Chapter raises question of crucial difference between "international" and "global."

This scaling up problem definition engineering is one way, in Gary's view, of creating global engineers and that's the whole point of the paper.

I would like a little bit more detail about the Virginia Tech program itself, the Engineering Cultures course.

I thought the piece on the distance learning was very interesting. I would suggest looking back at how that piece fits into the overall concept of what you're trying to do through Engineering Cultures and problem definition.

Joe raises in a very eloquent way the question of the need for engineers to have international experience. He's hesitant to accept that fact. I think it's important to take a look at this in more detail. [not really about Gary's ms].

One of the fascinating pieces of this whole project is for each of us to read people's contributions from really different points of views, with different writing styles and the like. I don't read very many papers by engineers, for example. The writing here is goal-directed but I don't know if I would change it.

There are a lot of questions raised. One technique for writing is to raise a lot of questions and to leave those up there for people to contemplate. That's a very good technique. Still, I would like a couple of answers.

Why do people in different cultures or some different backgrounds define problems differently?

I was really interested in this comment that if the core of engineering learning included defining the problem collaboratively with other, it could help engineers and educators see the practices of design, ethics, communication, leadership, diversity and impact analysis of internal not external.

I want to ask him how a successful STS program brings these kind of questions into the heart of the engineering community.

It is still not clear in my mind how much overlap there is in the "problem definition" aspect of engineering emphasized by Gary, and "open-ended problem solving" which normally involves customer or client input (e.g., voice of the customer) and formal steps related to problem definition. Certainly, the customer perspective starts out different than the engineer's perspective, and hence I continue to feel that problem definition in engineering is always part of "problem solving". I was intrigued by the phrase in the paper "Engineers, in this view, have been built to be servants of progress. Understanding this is important because it helps account for why engineering education appears to be in a perpetual state of reform." Being a servant to progress, broadly defined to go beyond technological progress, is a good thing in my judgment.

Now, I'm certainly not as familiar with the author's data as he is, but I must confess to a certain hesitation to embrace this theory wholesale. The fact is, there are certain indisputable facts embedded throughout engineering curricula; sometimes there really may be only one way to pose a problem, and certainly many problems have only one correct solution. The end product of an engineering education should be a person capable of creatively applying sound math and science to the betterment of society, and the creativity part of this definition can certainly mean different things in different places, and more importantly, it can be fostered in completely different philosophical approaches in different places based on local cultural and other influences.

In my own work with globally operating companies, I experienced a certain pressure due to efficiency issues to diminish, hide, deny, or downplay the fact that engineers from different cultures DO have different ways to frame a problem and pathways to solve it. There is a tension within those companies whether to accept the relative "messiness" of international projects, i.e., how to structure communication, management responsibilities, etc., or indeed try to promote a new "global" engineer who leaves many cultural traces behind – if that is actually possible.

I realize how crucial "location," "knowledge," and "desire" are. But what about "power," "institutional center vs. periphery," and business demands and interests?

Author response: I did not realize how much I was depriving readers of the content of my trajectory in describing only how I worked to scale up the Engineering Cultures course and not on how it emerged in the first place.

CONVERSATION WITHOUT GAYLE ELLIOTT

This manuscript offers a thoughtful tour through the details of managing and administering the International Engineering Program as it evolved into International Co-op Programs. The major contribution of the manuscript at present is to document the difficulties an institution faced in integrating "international" into its routine practices. The case stands out in that it is built around the system of co-operative education.

The division of Professional Practice does not have a counterpart at my university. I would like to know a little more about it. How many staff and faculty and what are the responsibilities?

Someone suggested this geography could benefit from more student testimonials, but beware that student testimonials can become self-serving.

Something that was mentioned but not elaborated is that the author has been involved in fund-raising. I haven't heard anything about fundraising in our group. I'm interested in how it fits into the struggles and barriers or successes.

I think the program right now is at twenty-four German placements and twelve Japanese placements. I wasn't quite sure where the Spanish program was at.

Here's an example where the support from the top left, yet the program still went on.

Is there any participation from the College of Engineering in this? There certainly should be. If there's not, I just wonder how endangered the program is. Is this program there because Gayle is there? Or is it institutionalized and secure and is valued from the president on down, including from colleagues in engineering?

I wanted to know more about the author and her career? She talked about her own international experience during high school with a trip to the Soviet Union. How has her own training and background shaped her career?

I would like to hear more about how she actually finds work placements in Germany or Japan. I don't think she gives herself full credit for that. How did she have the nerve to do that?

You talk here about how the Japanese and German companies want to expose their employees to American students and American culture and have them practice English language skills. That was nice because we often don't know what should compel a foreign company to accept American students when there's no reciprocal exchange.

Why would a partner in another country want to accept American students in co-op positions?

I found myself confused when reading this paper until I concluded that the author approached international engineering education not as an engineering educator, but as (I think) a higher education administrator with academic training in a field other than engineering. I think that readers would benefit from more information about the author's background so that they understand the perspective from which she approached her work. It's hard to tell if her increasing involvement in international engineering education was an unusual risk, a logical career move, or something else.

Author: I didn't go to college when I got out of high school. My Dad went in the Navy. My mom went to work in an office. If I had wanted to, they would have supported it. But I grew up with the expectation that I would not go to college, and so I didn't until 10 years after I started working at UC.

Author: I wanted to clarify that the program moved around the University from engineering to global studies to professional practice because of a desire to improve and expand it. Our engineers don't study abroad simply because between co-op and the engineering curriculum there's no flexibility to accommodate that.

CONVERSATION WITHOUT LESTER GERHARDT

Gerhardt's personal journey is a compelling description of his upbringing in NYC, his remarkable career as an engineer/educator/administrator, and his rationale for international education.

He had a career in industry, a research career and then became a college professor and then an administrator and then international educator. All these things were opportunities that were seized. I hope to hear more about that.

As one who has photos in his office of Gene Cernan, the Lunar Rover and Module on the plains at Taurus-Littrow (Apollo 17), and of "Earthrise" signed by Frank Borman (Apollo 8), I am certainly jealous of some of the experiences Prof. Gerhardt has had in his life.

I like the way Alan Parkinson said what he was looking for is the interaction of global activity and his life story.

The international internship is a different experience than taking courses, which is a different experience than a service experience. Each one of those is going to have a different learning outcome. Is it all equal?

I would love to know about the conversations, struggles, challenges, hurdles, aspirations, and visions behind RPI's requiring an international experience with each undergraduate student. Are the faculty going to be involved in this? How is this going to play out in the curriculum? What is the role of global experience in the modern engineering curriculum in the author's eyes?

Les has always challenged us with the notion that there are plenty of resources on our own campuses to provide students with international experiences.

The author should maybe think about how he is going to be a role model for others because his trajectory is very unique. I would say it would be neat for RPI students to read this and realize this guy up there in the administration took a path that was not too clear. It wasn't obvious that he was going to be here.

We each entered this discussion with our own internal sense of what's important about global learning. It's so internalized we're not used to talking about it. We may think it's obvious, but we have different senses of it. Some of those senses include a lot of learning about language and culture and some don't. Some are more about engineering as a professor, some about making the world a better place. We need to begin picking apart the why and the what. There's only so much that can be included in engineering education. All programs don't need to be the same, but they need to be somewhat intentional about what we're trying to achieve, what are the learning outcomes.

I'm struck by how the intellectual content of someone's work affects how the personal and professional intertwine. This guy got interested in visual simulation, which required seeing things from many points of view. To what extent might this interest in visual simulation have played a role in shaping these decisions?

Author response: Every student who has done this under any program whatsoever has come back with a life-changing experience. It doesn't matter where it is or what it was. Eventually I came to the conclusion that we as intellectual leaders and mentors to the next generation should be telling people to just do it.

Author response: I never did any self-reflection of this type. It seems I've lived my whole life without thinking about why and how I did it.

CONVERSATION WITHOUT JOHN GRANDIN

The great value of this chapter and the collection is the kinds of lessons and themes that come out. This trajectory started with a chance circumstance in a backyard, and it grew to become one of the granddaddies of international programs. This trajectory is really unique and interesting, especially for people who are just starting with that chance circumstance.

This program has been in existence before global engineering became popular. I think that's really great.

Significant is the total immersion aspect of the Program, not only by incorporating language (not unexpectedly given John's specialty), but by creating a separate dorm on campus with a German chef, etc., for truly a total immersion experience both on and off campus.

There were several comments on the significant amount of fund raising and not having an operating budget for twenty years. It's pretty significant he could maintain and grow that program.

To us, this is a very poor institution. Your achievements, including the amount of money you've been able to raise, that needs to be pointed out.

He begins to tell us how he challenges his colleagues in literary studies to show relevance to business, to engineering, to chemistry. That's huge when you take on that role. That's a huge political endeavor to challenge your colleagues in that way. That political work deserves more visibility in his geography.

Scholars in the humanities often disregard input from companies; they see corporate motives as highly instrumental. This boundary crossing must have been risky for him, especially if he did not have tenure at the time (this is not clear from his account).

It's clear in his narrative how important the dean of engineering was as a champion for his program. And the subsequent deans pretty much fell in line. Well, I'm sure that if they did fall in line it's because of John's agency in educating them and helping them make the case for his program within engineering.

Some deans and associate deans are asking for these things. But then there's very traditional faculty at the core of the curriculum who are powerful, sit on the curriculum committees and advise students but are not welcoming of these persons. So how did John help deans make a case for these programs?

How did he maintain relationships with his main benefactor and with Texas Instruments? This is not a trivial thing. To a large extent, the success and sustainability of these programs depend on these people. Sometimes they are very uncomfortable because you don't see eye to eye. Yet you need to find ways to maintain and grow these relationships in ways that are good for your program and your students.

This program started with German. How do the people who have shaped their identities around other languages co-exist around John? Is it a peaceful, harmonious co-existence or not?

I was very curious in reading the short description of the Chinese programs that are emerging. What are the funding sources? What are the motivations for the various stakeholders? What work does John do in determining mutually beneficial relationships? Has it been different in the Chinese case?

Is bigger always better? This is already a fairly large program. Part of the beauty of my own program is that we have a personal connection with all the students. If we scale up too large, that becomes an issue.

Many charismatic leadership projects never make it to permanent structure. So how much does this rely on the charismatic leadership of one person? The author needs to talk more about why after twenty years there is so little institutional support.

What's going to happen when he goes? What are the considerations at the end of the process, as he looks forward to maybe getting a life again? What is going on in his head? How is he talking to his senior administration about what comes next?

Author: Some people said I made it sound too easy. It hasn't been easy. It's been fun, but not easy. One book I thought of writing when I retire is called, "You can't do that." Because whenever you're crossing boundaries or doing things within a structure, someone is always there to say, "No, you can't do that." So you have to figure out how do you do it when they say you can't, and get it done.

CONVERSATION WITHOUT DAN HIRLEMAN

The description of the Purdue years and the development of GEARE is insightful and readers will appreciate the background information on this unique program. Interesting to note that the initial attempt at a large scale program did not succeed.

The author's description of his personal history of first entering graduate study in the US, then becoming increasingly involved in research, and finally having his research lead naturally into some international collaborations, illustrates a path that is fundamentally different than that of many contributors who are now involved in international education.

Paper does an excellent job in describing the current international global activities in Mechanical Engineering at Purdue. It also addresses crucial issues such as language learning and the increasing role of service learning opportunities for students.

As the author stated, he's focused on industry-related internships, but then he also toward the latter part of the chapter talks about service-learning projects and how students' motivation brought those projects into the programs at Purdue.

I'm fascinated by the author's work as department head. Why in the hell would he move from a career in research to the position of department head? His commitment to helping students certainly has dimensions of helping American competitiveness, but there's other stuff in there and I don't quite understand it yet.

We need to get students to become equally involved as Dan got involved within Denmark, creating structures so students don't wind up hanging out with only Americans.

It's quite interesting that the author pursued an endowed faculty chair as his first international activity as ME Head at Purdue. I'm not aware that such an initiative has ever been mentioned in the context of globalization of engineering education, discussion of which is commonly student-focused. It's a terrific idea!

Whoa! This is a huge innovation. Engineering curricula typically do not include minors. Might the introduction of minors be a key curricular pathway for integrating global studies into engineering curricula? What is an engineering major that includes minors?

Certainly the author makes very clear that Purdue is reaping the benefits of international experience that he acquired while still a Purdue student, and why shouldn't they expect even more of this from today's batch of students?

I think the key word in there was faculty ownership. I think sending students abroad to other programs does require faculty ownership, you know? That's one of the things we still go back and forth with at Georgia Tech.

One thing I really get from this manuscript is a developed picture of work designed to help students get to industry, work in industry, and make a difference in industry. For people who define their work as doing something other than that, there's a stereotype about producing students for industry. They think they understand what that's all about. It's not that at all. It's complex.

I want to disrupt the opposition between changing the world and going to industry. It sells industry short, or it may sell industry short. If my career is committed to life in industry, I believe I'm helping the world. I'm contributing, and it might be through increasing standards of living. Readers need to see that.

The point I'm trying to make is we need more accounts of life within capitalism. There are many capitalisms. There are many people who find meaning and commitment and fulfillment working in industry. I resist the urge to paint it in the singular.

The author is in a unique position to tell us about his partners overseas. What kinds of barriers did he have to overcome? What can students get by being exposed to different engineering cultures? Maybe different industry cultures?

I have this idea that when students with international experience get into industry, they develop different kinds of relationships with their partners abroad. If we're going to go to China, doesn't it make sense to make a connection?

We want to produce smart people who can endure complexity. That's what brings people forward in companies. Saving the world is fine, industrial competitiveness is fine, but there is more to it. What does it mean to be an educated person, an educated engineer?

What is the result of creating these globally competent students. Is it to save the world or do better work in your company? Students say this experience had no effect on my hiring, but once I was in the company, boy, my options were amazing. We hear that story over and over again.

Do we know that service learning projects have a positive impact on those who may benefit?

Author: I do have an industry bias. I don't know if bias is the right word. But certainly of Purdue undergraduate engineers 70% go to industry, and probably 70% of our Ph.D.s go to industry when they leave. So, certainly, industry is a customer if your want to use that word. So I bring that in.

Author response: I did treat this as an engineering design problem. You gotta define it. What's the problem? Who are the stakeholders? What are the constraints? The problem is I don't have any students doing this, so how do I get them to do it? How do I pay for it? So that's why industry. You gotta pay for it. So that was the vehicle, very pragmatic.

Author response: When I created the global engineering program, I remember my boss calling me and saying, "Are you sure you don't want to call it the 'international engineering program'?" And I was sure I didn't and made an explanation, but I don't know how convinced I was.

CONVERSATION WITHOUT JUAN LUCENA

This paper is a great addition to the collection of papers as it does not really discuss the inner workings of a particular program, but instead it discusses a concept. The points brought out in this paper are important and should help the workshop participants reflect deeply on their programs and on their own personal geographies.

Most of all, it speaks to the importance of the broad issue of the need for curriculum development to incorporate a global dimension as part of our teaching and learning programs on campus.

This should be required reading for all students.

I thought this was the most philosophical of all of the papers I got to read, and I enjoyed that very much. I didn't find it scholarly in the way that I normally think of scholarly works as much as I found it very intellectually stimulating. I thought it gave a certain depth that's very much in keeping with what we're trying to achieve with this collection.

The author indicates how engineering was not really an acceptable career path as seen by his family and community [in Colombia], for he was not a lawyer or a doctor.

It struck me that Juan avoided self-reflecting about his own pathway into engineering. So, I would push him, as I think Rick mentioned, to share more about what led him to the United States. What led him to attend RPI? And what led him into engineering?

The author develops the specific Engineering Cultures for the Developing World course. To what extent is it part of a larger scheme to prepare students for the global enterprise?

I've often wondered exactly what it means to produce "ABET-like engineers" everywhere. I still find it hard to believe that engineering is so cut and dry, one size fits all. Cultural diversity must somehow lead to entirely different ways to formulate problems, and therefore to solve them. But do the tools used to solve them necessarily also change?

Juan has walked the walk where most other people have talked the talk. He's experienced it. He does it from real-life experience. I found it really exciting, really valuable, and real most of all.

I'm not sure we all know as engineers how to disseminate things beyond the engineering community. I thought the author's discussion on the issue of disseminating something that's outside of our profession was really good. I'd like him to reflect on that a little bit more.

A healthy discussion is going on right now with the faculty in the college of engineering about general education, you know, social science courses or humanities courses to be taken along with the engineering program. The course like the Engineering Cultures course is extremely relevant to this. I would see it as a very important course for engineers to be taking and also a course that can be opened up to the entire academic community as a general education course with the role of technology and what engineers are and what engineers are for and how they are seen in France versus Columbia versus the Far East or whatever.

I think it's valuable to think about and to explore how this experience fits into the larger menu of opportunities that students have. Hopefully, it's opening windows for them not only in terms of their educational experience but down the road in industry as well.

This has been a somewhat difficult paper for me to review, as I have not been exposed before even to the ideas of "Engineering Cultures." I have to admit that until now I had not even considered that engineering education might be driven by political drivers such as the Cold War, or that engineering is fundamentally different depending on your country of origin.

I struggled some with relating to the author's struggles. Engineers are not always the most reflective people, particularly when it comes to reflecting on what they may perceive to be rather obtuse questions, such as, "What is engineering for?" Many engineers would feel the answer to that question is self-evident.

I also am not sure what he means by "neoliberal ideology," although I am reluctant to admit it.

Author response: The course now lives in the humanitarian engineering minor. Actually, the one that I teach is Engineering Cultures in the Developing World. I continually pose the question not only to myself but to my fellow engineering peers—What is engineering for? Unfortunately, many of them, who are particularly key actors in the humanitarian engineering minor, have taken that question to be offensive: "What do you mean, What is English for? What is engineering for? We're here to save humanity. We're here to save the poor of the world." Of course, I find the answer equally offensive. So right now we're in kind of a stalemate because no one wants to sit down at the table and say what exactly we are doing here.

I would argue that what you have experienced as a stalemate the students might experience as a very interesting tension between two different ways of looking at something. It gives them choices about how to think about it. So it's possible you could present both sides of this argument. Maybe just describe this as a challenge for anybody who wants to be involved in humanitarian engineering. They need to resolve these questions for themselves.

CONVERSATION WITHOUT PHIL MCKNIGHT

We follow here the pathway of a German language and literature researcher and teacher as he makes his way into language education for internships and ultimately into developing this massively impressive international plan at Georgia Tech. Along the way, we learn how faculty in foreign language departments are often unwilling participants in such collaborations despite clear expressions of interest from the Modern Language Association.

This is an account where it seems like international education saves the author in some sense.

In the 1960s, this guy organized a symposium on *Why do Germanics?* He's been carrying these questions around for a while. At first he says, "Screw it, I'm moving on." But then later, it's, "Well wait. I've put all this time in. I have something to offer."

There's a lot of great material in this piece looking at the relationship between life inside the academy and outside. We get an account of his movement away from the academy, working in a motorcycle shop, and living life as a single parent. This interest in the relationship between inside and outside dates back to the 60s.

This is an interesting account written in a style quite different from most of the other manuscripts, especially the forays into much deeper philosophical dimensions (personal and institutional).

The "big-picture" issues of what exactly constitutes our modern educational goals in engineering, liberal arts, and languages, are complex; their ties to internationalization of engineering education are inextricably linked but barely understood. The author provides some excellent motivation to pursue this more deeply.

The author poses an excellent version of this question: Is the social responsibility of those who create and understand technology greater or less than those who for whatever reason, won't or can't understand it? Do people who, like the vast majority of modern Americans, refuse to pursue even the most basic math and science literacy in a technology-driven world, nevertheless retain the right to expect those who do to always act in and take responsibility for the public interest? This latter question also strikes at the heart of the concept of a "liberal education" in the context of a society totally dependent on very complex and ever-growing technology.

We're looking at something bigger here. It's about general education. U.S. engineering education has a general ed component, unlike many, many countries. What do we do with that? How do we best use that time?

Few pleasures in life exceed the satisfaction of feeling like one has mastered a tough philosophical issue, but that feeling won't pay the rent, and it certainly is of limited interest to business leaders in need of real help or academic administrators in need of new opportunities to develop pioneering programs of high demand. Well done!

"If the principal 'employer' of German professors [Goethe scholarship] admonished us to get rid 'of this pile of books' and find out 'what the real essence that held the world together' was, why were we not taking his advice?" That quote really blew me away.

I appreciate this story because I am ineffective in grappling with the problem of relevance in the liberal arts. I'm not effective because I become very confrontational.

Again, we're not just teaching language skills. We're teaching culture. That has to be brought out a little more in this piece.

Mike and Bernd in particular appreciated him calling attention to the disciplinary conservatism that lives in fear of the image of a service department.

Phil makes the important point that the MLA is a conservative organization. I've heard in public statements made by senior people in MLA that engineering is not the answer to reforming language education. They see it as becoming a service to engineering rather than some integrative approach to applying language learning.

Crisis spurs innovation. In some ways, the German field is in crisis and therefore seems to be innovating. Maybe you could explain a little more about that?

I'm interested in the modern language people because in my case the people that have supported me are those people. It was the flip side for me. That part is surprising to me.

Just to give rough numbers. In K-12 in the State of Massachusetts, 500,000 students learn Spanish, 40,000 learn French, and 2,800 learn German. At the same time, we as German professors are aware that what Germany has to offer is an incredible engineering culture and tradition. Germany continues to be the number one export country when it comes to goods.

Georgia Tech is kind of unique in the enormous steps that have been taken. It was from the top down. It was a decision by a president associated with quite a bit of money to do this, and then there were capable people who came together and turned it into reality. The relationship they have with their institution generally is different than what Professor Widdig experienced at MIT. Probably those faculty felt their primary commitment was to their intellectual and scholarly field and not to their colleagues across campus.

What's nice about this story in my mind is that it didn't begin with a business-oriented approach. This isn't someone who just blindly said he did it because it was important to business.

Anu found this manuscript to be a great place for studying two different views of global competency. One involves being attuned to the impacts of engineering projects. The other is about building student resumes.

Author response: Students who have joined the international plan have significantly higher scores in the categories of wanting to influence the political structure, contribute to community action, keep up-to-date with political affairs, become an authority in my field, influence social values, help promote racial understanding, become a community leader, and develop a meaningful philosophy of life.

Author response: There's been a failure of the humanities to really deliver what they should be delivering. It would be good if we can bring engineering and the humanities together so there's a greater influence on public policy and understanding of how other cultures think about things.

CONVERSATION WITHOUT JIM MIHELCIC

This manuscript powerfully describes development of a master's international program in engineering. The core of the account is the recounting of issues Jim has faced in running the program.

The paper describes a "lone wolf" who has undertaken a huge project at personal sacrifice, but who has also reaped great personal reward through the outcomes.

How did he do this? How did he suffer? At what point did he get scared? At what point did he say, the hell with it, I'm going for it?

This program is unique because it's about graduate students. This really gets at the heart of engineering as a human endeavor because it helps students understand that problems are deeply embedded in context and have to be solved under constraints, sometimes severe constraints. You have to leverage local knowledge and indigenous capacities in order to come up with sustainable solutions.

It began before the emergence of things like Engineers for a Sustainable World and Engineers Without Borders. Jim was developing this program as sustainability was emerging as an organizing concept in civil and environmental engineering especially.

He makes it quite clear that younger faculty assume considerable professional risk by getting involved with such programs, and that even he, as a senior faculty member, is viewed with suspicion and doubt by many of his colleagues.

You had an enlightening moment in your paper when you said, "We should do that." Many people say that. Most people don't. You did. What is it about your background, your makeup, your personal set of convictions that aligned you with the situation and made you simply plunge forward with incredible investment of time and effort?

How is it that the program did not fit into the institutional culture [at Michigan Tech] because a lot of what we're all wrestling with is institutional culture.

James might want to describe how he came to the realization that the societal aspects were missing from engineering. What was he reading? What kind of experiences, people, etc., led to this realization?

After his early realizations, James begins to cross boundaries, meeting with social science faculty, reading outside his field, and moving from lab to community. What did this boundary crossing mean to him? What did he learn about himself?

Maybe what Jim is doing is bringing education back to what the profession is all about.

I'd like to hear how has this impacted the way he relates to his field and his students, and even on how he lives his life

We do about as poor a job as humanly possible by having the world on our campus and not utilizing that resource. We all have once a year a food thing in the evening where they have food tables from their countries. But most people don't have an on-campus international experience with the abundance of the world right there.

Graduate students cannot automatically become global through exposure to international students because it doesn't remove them from their comfort zone. Nothing can replace hands-on experience in another country.

What is the appropriate level academically for global experience? Sometimes you have students going into a situation with completely unrealistic expectations. And you have faculty who have to balance the learning with what is also for the good of the project. It's really hard to measure how successful both sides of that coin are.

I'm so frustrated by always having to explain why we should be doing this. Not how and the details, but why is this important?

To get back to your graduate vs. undergraduate thing, in my experience it's harder to get graduate students international opportunities because the faculty don't want them to leave the lab.

The biggest difficulty I have had is with international engineering faculty. I went to see professor so and so, and they say, "Why would you go there? I came here."

I loved the story of the duck. I'd really like to know if he got reimbursed for the duck.

When we're doing programs like this, we're continually bumping up against the status quo.

The manuscript risks characterizing that pathway as not doing the right thing. Is that how he sees it?

Author: One thing about the partnership. It's not about a partnership with the Peace Corps. It's about a partnership with a community-based organization and the Peace Corps is not the only place to do it.

CONVERSATION WITHOUT JOE MOOK

This paper gives an interesting history of the (mostly self) education of an educator in the domain of International Education. It also gives some advice to faculty who may be asked to take on administrative positions related to International Education.

The narrative provides an honest account of how a "typical" US engineer developed an interest in international education. This "out of the blue" path is more common than we think, and it's important to note that it does not take a lifetime of international travel or fluency in foreign languages to make significant contribution to IE.

It's an interesting story because in the end what we find is there wasn't the kind of support that he describes is needed to sustain an effort.

It was really interesting that the initial conversation with the dean was, "We kinda knew we have to do something in international programs. We knew it was important and we said, 'whatever exactly that meant'" (quoting Joe). And I think that was a nice way to lead into the exploration of, *Here's this new job, what am I going to do?*

At this very early point in time, how did he understand what he was trying to create?

Using outgoing undergraduates to create tuition waivers for incoming graduate students was genius!

I'm intrigued by this idea that he never had any interest at all in international education and then suddenly something happened. Something must have happened in Hanover. And I think that the life-changing experiences that clearly led to the involvement with all of these other projects, I think that is the origin of this.

What does this shift in specialty mean in engineering terms?

I was really intrigued about his partners. What's the culture of the partner in the EU or in Indonesia and how did those partnerships work out. So many of our projects depend on partners. Who is a good partner and who is not a good partner?

In sum, this is a very complete and useful account yet I would like to know the history, tensions, and struggles in getting to each destination and the social and political networks surrounding each of these.

It seems from the narrative that language learning did not play a big role in the IE. Why not?

The author raised great questions on the time element…As a group, it would be very worthwhile discussing what is a minimum criteria (time and other parameters) to have a meaningful IE or GEE experience on BOTH sides.

How applicable are his suggestions to a variety of different kinds of institutions?

I'd like to know more about why you think the new Dean eliminated the associate dean for IE position. What does this mean for IE at UB? But even more importantly, what does this mean for your career?

Author response: Our office of international education is almost entirely focused on money-making activities for the university. I came to see that as least interesting to me.

Author response: I think most of us are here because of our experiences trying to create global engineers and do global education. I couldn't have done it if I didn't have the Global Engineering Education Exchange. Going to the annual meetings and hearing from thirty or so other US universities was crucial…I don't know that my situation carries over in the same way a lot of others do, because what happened is we got a new dean.

CONVERSATION WITHOUT MIKE NUGENT

To hear this from a program manager is both refreshing and disconcerting – refreshing that the problem is acknowledged, disconcerting that the ambiguity exists within the agencies responsible for deciding how to invest in the future of the field, and not just in the wildly-varying academic community's various schemes. Excellent food for thought!

This was a really fascinating reading. Many of us in academia are not aware of the constraints and structural challenges in funding agencies.

A key feature in this account, especially toward the end, is that it draws on interviews with a handful of project directors to document their perspectives on participating in the program.

Another important concept highlighted in this paper is the idea that global educational experiences must be integral to a curriculum (as opposed to add-on) in order to be broadly effective. This is true both from the point of view of students who need this to see the centrality to education, and to engage the faculty.

FIPSE programs were designed to create *meaningful* time overseas. And that is a loaded word. But we've all used it. And we know what *isn't* that, but to define it, we know it when we see it. But to put words to it is harder.

I had the pleasure of meeting this author yesterday, and one of the first things he said is, "I'm not an engineer." But every single time I got to talk to him he was really passionate about this stuff. So I want to know more, specific examples, about that passion. What brought him to his place in his journey?

What made him decide to go into government administration? I would also like to know more about his role within this organization. Sometimes our preconceptions about Washington are that it's just another bureaucracy. Are these all bureaucrats or do these people have similar passion as Mike has? I found it fascinating reading because we know so little about what's going on in those agencies internally.

I noted this course, Spanish for Engineers, which is interesting. It's ironical in some way right? That you have to create a course to make it work just for engineers.

How do faculty abroad assign grades? For one thing, they can be more subjective, especially if it's an oral exam. I know we have had discussions with a couple of people at other universities about whether or not the faculty member will take into consideration the fact that the student's German is not the same as other students there. Mike would be the right person to go into some details on this issue.

Those negotiations I found really interesting. Some of the rules and regulation they have that you can only do something if you work with a myriad of other universities I find to be a recipe for failure. It becomes incredibly difficult to maneuver dialogue where people across continents and different universities are involved. Who makes those decisions that are so far removed from the way things actually work?

How do things work in partnerships with other countries? How much was personal and how much institutional?

Do you have a sense of their outcomes assessment from their side? What these other countries are looking for and what they are finding and how that is matching? I think maybe in these partnerships it's a personal relationship that's key. That's been my impression. These are personal relationships. If the personal fails, the partnership fails.

In the Global E3, we have twice as many students going to Holland as to Germany. We have more students going to Denmark than to Germany. Why? Because in those small European countries, they do not assume anybody would learn Dutch or Danish. So the smaller countries are avenues for students to go to Europe without a strong foreign language background and don't want to go to the UK. So how do we deal with this. Someone needs to learn German to go to Germany, but anybody can go to Holland or Denmark or Sweden or Finland or Norway without learning a foreign language.

What are the limitations that occur [in programs like FIPSE] as a result of political directives? If you go all the way back to after World War I, foreign language study was forbidden in this country for a while. Then with Sputnik there was a huge comeback. There was the National Defense Education Act. Then for this long period of time there been a lack of interest. With the recent situation that led to the war in Iraq, there's been a renewed spike in learning critical languages, including Chinese. There's a political element that drives this.

Even in places such as France, teachers want many, many sections in English. Valencia teaches in English.

How do you do a personal geography in a way that has critical distance but without attacking? I would like to see that here.

I recall having been approached several years ago to participate in one of these grants targeted at Europe. After discussions with a colleague in International Affairs, we decided to decline. The principle reason was the sparse amount of funding that would have been made available. We didn't think it met the effort required to accomplish the goals. I would like to hear more on why FIPSE decided to spread this grant in what I would describe as much too thinly.

Author response: We had a huge amount of autonomy. We could make decisions internally on what grants got funded, based on peer review. But it wasn't just peer review, which is very unique.

Author response: On the political side dealing with other countries, it really was a personal thing. The guy that stood up in that hall was talking directly to his doctoral advisor, so his passionate speech convinced his doctoral advisor.

Author response: As far as the European Credit Transfer System, that is another book in itself. The Europeans basically gave money to the European programs to force them to use ECTS. It didn't work. What we did was say, "You come up with your own arrangements, and you have a year to do it." That worked. So the solution was to develop individual agreements before they start doing it, and until they got that going we wouldn't give them funding. FIPSE's primary principle is, *we're not giving money out for things that occur already. We're going to give money for something that's different.*

CONVERSATION WITHOUT ALAN PARKINSON

We have the perspective of a dean here, as well as of someone who is at a unique institution. All the institutions we've been talking about are unique, of course, but BYU stands out in particular ways as being a very fertile context for opening up global engineering education. Alan speaks to some of those enabling factors that are sort of unique to BYU as well as some of the struggles.

I wanted to comment that I really appreciate the diversity of people that are in this group. He is included as an administrator. I don't think I could ever relate to how an administrator thinks, so it was really interesting to me to hear his perspective.

Alan doesn't talk about his own missionary experience and how that maybe did or did not have a transformative effect in his life. He also did not talk about other personal motivations.

Phil raised the question about his missionary experience in Japan. I would like to hear more about the whole LDS belief system and how they see it interacting with the idea of setting up study-abroad programs and service learning things. I just have the feeling he's so disciplined and organizational in his thinking that he might find it difficult to delve into the biographical details.

Many want to hear more about the struggles and uncertainty. There's the LDS, your know, the church. There's BYU as an institution. There's the college of engineering. There's Alan's own personal perspectives and views. There's industry and donors who are involved in this. There's BYU students who are changing. What's a millennial BYU student? What are they looking for? ABET accreditation criteria. I think it's really interesting BYU is partnering with institutions like Georgia Tech. How did that come about and how is that working?

It is for him a pretty much done deal that the global engineer and the Mormon mission go together very easily? Or does that create tension and difficulties for him?

Coming from a faith-based institution myself and knowing that I am of that same faith and have similar experiences, I'm really interested, and maybe it can help me with my paper to see how he could weave those two together a little bit better.

My son goes to a church preschool and when you leave there's a sign at their exit that says you are entering the mission field. I love that. I almost think we need those signs on the doorways exiting our classrooms. We want students to be activists, to be change agents. Maybe there's some insight to be gained from that.

He [Alan] came and started with us [the Global Colloquia], asked me [John Grandin] for a few papers, contacts with other people. Then he himself published a paper saying this is what is going on. A year later he and his associate dean were reporting at ASEE about the programs they had set up. I wish we had a few more deans in the country that were implementing things as rapidly.

I would just like to add on I think that BYU students are more culturally aware leaders, you know, that honor integrity. So I think he has a challenge here because his message, or the themes of his geography, will appeal to a lot of us who are engaged with students that have some faith base for why they want to do some, have a global experience, and a lot of us interact with students that have these qualities because they are in an honor's program or that is the kind of student we attract. So he has a challenge here to make sure that his geography appeals to both of those groups.

Gary pointed out ways in which maybe some of Alan's engineering training and his business training are coming into his approach to developing these initiatives. I especially liked thinking about this interest in optimization. Alan is performing some optimization, which is very similar to Dan performing a design-oriented approach to tackling these challenges.

I would like him to take up the issue of scale. Could we scale up more using as a foundation the unique qualities of students that are going to study abroad or can we not scale up because the only students who are going to want a global experience are already leaders? So that's going to limit them, and we're going to stay at one percent or three percent the rest of our careers?

I found it quite interesting that he found in business, in an MBA, the vehicle he wanted to enact institutional change. He had to go to the business world to learn about strategic planning and try to enact change in higher education. I find that very seductive. Then, of course, you go to the reality of the curriculum committees and the faculty meetings, and that's a completely different reality, right? I don't know how comfortable he'll feel, but it would be great if he would just feel completely at ease to be honest and clear and transparent about this.

It would be nice to see some of the complexities of life in the dean's chair. People are saying, "Why the hell doesn't the dean just take this step, you know?" Well, deans and provosts and presidents have lots of developed constraints. The dean says, "Yes this is a noble and important effort, just like the eleven others that have come to me. All are important efforts and deserve funding. I have to figure out which to fund."

Something I want to say about my own paper as well as this one—knowing this could eventually be published puts certain constraints on what I do want to say about the political processes that are going on. I could, for example, say some really nasty things about my president. Do I want to do that?

There are some things you can't talk about publicly. But sometimes if you put in the effort into trying to figure out how that other side thinks, maybe you can distinguish between the guy who is taking some principled position here and the guy who is mad at me. Some things really are personal.

I felt the LIGHT acronym was "enlightening" but also was typical of such university initiatives, where there is more emphasis on the "words of marketing" than on the academic support and focus.

Alan starts talking about how competitiveness really is such a driver and yet he's had this growing awareness and growing interest in these humanitarian dimensions. A lot of folks really keyed in on that.

At the end of his account, he labels his projects "humanitarian" for the time. How has he dealt with the transformation of the programs from study abroad to humanitarian? Most of the training, resources, and points are not related to humanitarian endeavors, but to community development. Then he tells us that he came "to understand that another motivation for promoting international technical experience relates to the range and scale of technological needs of humankind in the 21st century." How did he come to this realization? Did he stumble upon the Millennium Goals? How does one begin to comprehend the "technological needs of humankind?"

CONVERSATION WITHOUT LINDA PHILLIPS

The story of a part-time, course-by-course, non-tenure track instructor creating and running such a program is very interesting. The related discussion of administrative support (lack thereof), research/tenure issues for regular full-time faculty, reluctance of deans and others to invest in this kind of activity, etc., are critical on many US campuses, and the perspective offered by this author is a great contribution to understanding these problems.

I find it important that we talk about power. Throughout her paper, there is the continuous worry over whether some higher authority will get the money and will I have this job next year, even though the demand from students was clearly there. It is clear from the number of narratives about students that how it's changed their lives is absolutely crucial to the author.

We have in this group those who had an idea and went with it and overcame all kinds of hurdles and those, like Joe and Linda, who originally start with "I was being asked", the whole premise is that things were brought to me and I responded. And at the same time in the narratives of those two people, at some point being asked suddenly switches to ownership and switches to this is becoming my project and actually this is changing my life.

What possesses a person to make this move and accept what is obviously a marginal position in the department? It must be linked in part to family missionary experience but certainly also to what it meant to work as a construction engineer. I'd like to see a whole lot more on life as a construction engineer.

The connection to some fortuitous ABET changes is also quite interesting and brings in the element of "random" historical timing.

This makes me aware that it's so different if someone does an internship with Siemens, a service-learning project, or an experiential project in another country. The author addresses very well the issues when one engages young engineers in the complexities of the developing world. I would like to learn more about how the political context may have changed [in Bolivia with the recent presidential election] and new contacts had to be established.

A key point is made on the final page regarding the education of socially conscious engineers and the need to cross traditional disciplinary boundaries and achieve diversity in the field.

Most interesting was the described reaction of professors (who brought the students to tears with their comments) and the members of the industry advisory board, who unequivocally stated they would prefer to hire exactly these kinds of students.

We know that female and underrepresented students are more and more looking at engineering when they are able to see engineering in a non-traditional sense. Service learning appears to attract a lot of female students into the service projects because they see the human value involved in that.

It was just a two-week program in Bolivia. How could this have had such an overwhelming impact on these students, as evidenced by their comments? I'm really stunned. I'm assuming it was the preparation by the author.

How much did having a certificate on sustainable development help prepare the students before they left?

I would like to see some discussion of sustainability not only in this paper. We keep seeing over and over again this question of whether these programs are going to live and die with their founders. Is it just all about a bunch of people working seventy hours a week?

Why is the university not giving this program more resources?

The paper as written has a bit of an "us vs. them" flavor, i.e., the course champions vs. the rest of the university. There are not many insights as to what might compel the university to scale this up and support it financially at a more appropriate level,

Aside from the potentially thorny "legal" issue of mixing religion with a public university program (described in the paper), does reliance on a missionary organization tend to isolate the students from more direct contact with the local population at large?

The question is whether the church represents the mainstream indigenous religion/culture in Bolivia, or if it has a mission to proselytize, since such a mission would be inconsistent with cross-cultural respect we are hoping to promote.

The author points out that it is valuable for students (and professors) to recognize that technical challenges are closely linked with social, political, and cultural factors, and, I am sure we would all agree with this statement. Yet, there is little narrative provided on what prior training the teams received to learn both the theory and techniques in these non-technical arenas.

How much of what the author does depends upon trust with the locals?

What isn't clear to me is how much actual local interaction occurs in defining and meeting these local challenges.

Figure 4 appears to assume that clients *will* approve the design – where is the place for the community to critique the design? Should the benefactor and the clients be separated in interviews and group meetings, so the clients can express their opinions freely? Similarly, what about gender issues?

Given that we in US academic institutions reap the greatest benefits from these projects, what sort of standards are we willing to set as a professional community on many of these important ambiguities – choice of partners, prior training of students/mentors, choice of mentors, outcomes assessment (not just for US students but for the communities, conducted in an objective manner), etc.

Author: I see myself as a transition, to take students from those calculation and theory classes to the practical, to that next step of going out the door into industry. Being able to take somebody out of their comfort zone hits you someplace. So that's where I am and pretty much my focus. I think the International Senior Design program has been fairly successful…But honestly, nobody in the hall talks to me or to Dennis. We're different and they don't know how to think about us.

CONVERSATION WITHOUT MARGIE PINNELL

For struggles and such, I think this chapter was especially strong from an assistant professor's view-point of what is encountered along the road, especially in terms of how the reality of tenure may be different than the stated mission of a university. The chapter was one of the most personal ones I reviewed.

She tells us about her years as a stay-at-home mom and the skills and values acquired during this period of time and how they carried over to her current career and as a faculty member in engineering. Skills like management, multi-tasking, creative problem solving, and so forth.

This program and this work has enabled her to unite her personal values with her teaching, research, and service.

This is essentially a faculty member talking about her involvement in a program that was really a grassroots student movement. That's really unique.

On the negative side, the biggest disadvantage, as she calls it, is that her time spent on pedagogical projects such as this may not be rewardable in the tenure process.

We need another category of annual evaluations called global initiatives, global involvement, global whatever word you want to use. It speaks to the value that we're all trying to promote here. It would speak to the university and how it values our work.

What is it about us that when the opportunity comes along we lunge for it whereas others would let it pass by? Why does she feel strongly enough about this to put her career at risk?

Mention was made [that] the service learning included structured community service and structured reflection. The structured reflection was for me the most important feedback mechanism. All of us should be concerned about feedback no matter what we do.

I'm fascinated by this generation of students called the "millenniums." They do seem to want as a generation to work together, to make the world a better place, to help in some sense, and the service learning programs are aligned with that.

A huge research question is the extent to which Engineering Criteria 2000 is helping to transform the standard engineering curriculum and opening up these new types of programs.

I think engineers have a social responsibility to not only do the work but to understand the impact of the work, its effects. They also have a responsibility to orient their work so as to understand the world and make it a better place.

I like the cultural sensitivity class because I think that's something all students need. I would like to hear a little more about that course—the topics that were included and how they decided on that structure.

What kind of difficulties were encountered in making arrangements with international partners?

I have many, many colleagues who talk about how much energy and how much creativity they get from students. There's this recognition that a lot of what we like to do as scholars and as educators gain meaning by jumping in there with the students.

Later, we find direct and unquestioned assertions about how science and engineering improve the lives of communities. Has Margaret questioned any of these assumptions about "help," "needs," and of course the inevitability of science and technology as saviors of the world's poor. Finally, she makes an honest confession that "the ETHOS program saved her." Quite often this is what these type of programs end up doing (i.e., saving our conscience; making us feel altruistic, etc.) regardless of whether we can show if they have a positive impact on the communities they are supposed to serve. Has she struggled with this question?

Author response: I made a decision early in my career I was going to do things I felt were good for the university, good for the students, and good for society. If they didn't give me tenure, then I probably wasn't a good fit for my university.... I didn't start stressing about it until I started putting my tenure portfolio together and realized that I really wanted to be at this university.

CONVERSATION WITHOUT ANU RAMASWAMI

This paper is a radical departure from all others. It argues that global education should start and be exercised on the local level.

This little title that "The answer is here" is so provocative. That really is something that I think almost no one else says this about global engineering. It's just assumed to take place outside our borders somewhere.

The author provides a vivid description of her childhood in India, the roles of women, and the value placed on educational achievement. These influences came together to shape her view of engineering work and career path.

This manuscript is a delicious tease. It traces a trajectory from India to the U.S., back to India, and then back to the U.S. As such, it beautifully calls attention to the increasingly ambiguous identities of engineers in the present.

The text does delve into questions regarding the relationship between the developed and developing world. In particular, it is pointed out that the developing world has practiced sustainable engineering for centuries, without much benefit from technological advances of the last century. So the notion of the developing world going back to help with sustainability does include a degree of irony.

I want to emphasize that the issue of the mother and grandmother being food science engineers and biological and chemical engineers is really good.

If the author had stayed in India, would she have become the kind of engineer that her mother and grandmother were rather than the kind of engineer she became?

Whether foreign language is involved or not, a lot of what we do involves going someplace to show them how to do something. But this mother and grandmother already know how to do that and in an efficient way. Considering certain issues that have occurred in recent years involving energy and climate change and the environment, I'm wondering if the transfer can't go in both directions? I tend to think we rarely think along those lines.

A lot of schools spend a lot of international effort on service projects typically in the developing world. The mismatch maybe between what we are providing as American educational institutions and what's being received is important. The author has the advantage of having seen it from both sides maybe and has a better view of how this is perceived on the receiving side.

I would like to learn more about the Indian Institutes of Technology.

Can we really get a perspective on ourselves without stepping outside? People have blinders on. Is it in fact possible to get undergraduates conscious of other ways of thinking and other ways of seeing the world without putting them in a situation where they actually have to do that?

The international experience is like a lab experience, hands-on, part of the transition from this very classical, mathematical education.

What is perceived to be a good scale at U Colorado Denver? Is it expected to grow?

An interesting comment was the isolation within an Indian subculture in graduate school in Pittsburgh. We have around a thousand Indian students. It's certainly enough that there could be an Indian subculture and Indian students can choose to remain within it.

Are we kidding ourselves over all those service projects, parachuting kids for a couple of weeks somewhere into the world and saying, wow, we do good? These are very serious questions people in international education have. The author is in a particularly important position because she comes from India. She actually has a sense of the complexity and maybe the tremendous distance between an undergrad from Colorado and rural community India.

How do you teach an engineering scientist about India? What would the author think is adequate preparation for a student to spend four weeks or four months in a rural part of India?

Some international experience can be actually negative. You take them over for six weeks of drinking and they have no interaction with the population, so that has almost a negative effect on the students. At what point does an international experience actually give students that, "I get it now. I can see other perspectives. I can't use the usual answers here because it's a different context?"

A lot of global engineering in companies is done every day, and people do not leave their country. They use video conferences, teleconferencing, email, telephone calls, maybe once in a while a meeting at an airport. That is how most of global engineering is done nowadays. How are we preparing our students for that?

Author: When I'm thinking of global engineering, I'm thinking of the ability of our engineers to solve problems on a global scale. That doesn't mean making products with another country somewhere else. So I'm not completely convinced that you need to go all over to see another culture…

CONVERSATION WITHOUT RICK VAZ

The researcher's account is fascinating and challenging because his story is of an engineer who began with a traditional career in electrical engineering and little by little began making the move into this terrain he calls interdisciplinary studies. That jump is fascinating because he plays with this delicate yet unquestioned boundary between the technical and the non-technical.

The personal story of Rick's involvement follows a similar path to that of many students, but it is fascinating in that it is the story of a faculty member.

The program is often cited as a good example of international engineering education, and it is enlightening to gain some historical perspective from a faculty member who has been with it since its early stages.

He ended the paper with some really good suggestions, very personal and practical suggestions on how these programs work.

What's really unique is about migrating from a conventional educator and researcher to the Dean's office. It is a powerful story and lessons. I think a lot of people could probably see themselves in some part of your journey.

Here I was a recently tenured faculty member in electrical engineering. The question is, what do I do now? This opportunity came along and grabbed you in some way. That's an interesting phenomenon that a lot of us have in common. You called it a personal scale-up. I think that's an appropriate term.

Why did this grab you and not someone else? Why would it be of no interest to one person and yet you have this one experience, you hear other people's perspectives on things and boom! your career is taking a new direction.

Most of the people in this room are faculty based. Most of the meetings I go to on globalization and internationalization are frequented by people who are not faculty based. The role faculty play as mentors is key to the success of any educational program.

There's nothing mentioned about scholarship. Faculty are going to have to integrate their education with their scholarship so we can deal with issues of tenure and promotion.

I think the thing that makes us successful is our inability to accept no for an answer. You have an idea and instead of backing down when you hit the first obstacle you find a way to go around and keep it going.

These programs attract students who are broader than the stereotypical engineer who wants to sit in a cubicle and work on a computer.

The only little piece that was missing was it appeared so easy. There's a lot of work that goes into creating those programs.

Is it true WPI has no curriculum?

Faculty members talk about this as life-changing but don't really elaborate what's life-changing about it.

The issue I would address is sustainability. How does one assure continued leadership? How do you assure it continues after you leave? How do you convey the passion to the person who must carry it forward?

Who's benefitting from these programs? I would challenge Rick to speak in some way to an evaluation component. Are there any follow-up studies in terms of success rates?

Engineers like Maurice Albertson, one of the founding fathers of the Peace Corps have become quite critical of development projects. They seem to be more about the students and less about the communities we're supposed to serve. Very seldom do we take the time to assess the effectiveness of the project. Has Richard encountered these critiques, and what has been his reaction?

Given the level of resources which must be committed to this program, I would assume there were and are those who think they could be used productively in other ways. How real was that struggle and how real is it today? Does the program feel fully institutionalized and, in a sense, untouchable?

What makes this project global? What does global engineering education mean to Richard?

Author: Global learning is the platform through which we achieve critical thinking with problem solving, because we want students to understand they are solving problems in social and cultural context. I'm very aware of critiques of development and realize I need to say more about that and about assessment.

CONVERSATION WITHOUT BERND WIDDIG

The author's evolution of perspective with respect to two key issues—first, the role of foreign language and culture instruction in modern times, and second, the role of foreign language and culture instruction in an institution with a very strong science and engineering emphasis—issues which are not necessarily separate, but neither are they necessarily intertwined—is really fascinating and illuminative.

The paper also touches on a question which might be worded as: What constitutes an early 21st century liberal arts education?

Bernd is not only a contributor to the discussion here. He's also a person who has experienced some of what we're involved in. He experienced self-realization by changing culture. In his case, he stayed in the new culture and developed a tremendous success story.

So, you see this journey between two cultures and two disciplines and how the perspectives changed in his transition from Germany into the U.S. becoming the sunny boy down in California after coming from dark and dank Bonn.

This personal geography says something about the particular way this foreign language and literatures department was oriented more towards a traditional scholarship than it was to its institutional context. It is located totally in contrast to Georgia Tech.

Where are the engineering faculty in this discussion because they seem to be missing? I tend to think there was some elitism in the engineering faculty. I'd like to know more about your struggles in view of the engineering attitude toward these projects.

Why no study abroad at MIT? The whole program consists of internships abroad.

Can the MIT model work elsewhere, or is this exclusive to MIT?

I have a little concern about why these kids came back and said the foreign language deal was a waste of time because everybody spoke English with them. What can be done to fix that?

What was the role of foreign languages in and of itself? What is the role of foreign languages in engineering and science?

A number of reviewers would like to understand more about the internal conflict between getting his own scholarship done as an untenured professor under pressure and yet pursuing this entrepreneurial project.

I thought it was really significant the author was willing to take steps that were clearly not in his best disciplinary career interest for tenure and promotion. In the end, it turned into a negative tenure decision. Was there something he might have done that would have changed the outcome? If not, you know, what's the advice for the followers? He's made a huge contribution to promoting global engineering, but he did it by sacrificing himself. That's not something we can expect. So if our mission is to keep promoting global engineering, how do we get other language instructors to contribute? They are shooting themselves in the foot when they do so.

For me tenure is no big deal, because in industry we don't have it. What's important for me here is to see how it hurts some but I can move on. I can survive that disappointment.

There was some disconnect between him and his colleagues in Foreign Languages and Literatures because he worked with engineers. You know, those guys are the enemies. Don't be consorting with them. But he did anyway.

The initiative for these kinds of interdisciplinary efforts always seems to come from engineering. That disturbs me. Why isn't it the foreign language department or the humanities taking these kinds of initiatives?

What self-respecting professor would ever admit that a non-major student taking a course or two within the professor's specialty has learned "enough" about it? And yet the author's view of the evolution of foreign language and culture instruction in the context of globalization of science and engineering education is suggesting just this, which shows remarkable courage.

There are a number of people here from the German field. It's clear enrollments have been dropping for twenty years except in programs where they've done things like Phil, John, and Bernd have done.

Did the engineers at MIT already perceive the foreign language department as a service unit? Were the foreign language faculty saying, "We don't need to go there because then we'll be just a service unit?"

The engineering engagement with the language department often really is at the service level. They're looking for the first year or first two years of language training. So I completely understand why a foreign language teacher would be disinterested in being engaged at that minimal level.

One question that comes up here is what does global competency mean and what role does language play in that?

This man understood the tenure challenge at the point of arrival or at least he figured it out after a while. Why did he continue? Why was this so important? It seems to me it's related to the intertwining of the personal and the professional. He couldn't just reflect on the literature stuff. It wasn't right.

Author: For me, the personal and the professional are completely interwoven. They are not separate. I use the example of the religious order. I still feel, to some degree, I carry that attitude towards my job. I have been living in ivory towers all my life.

Authors' Biographies

KACEY BEDDOES

Kacey Beddoes is a Ph.D. student in Science and Technology Studies at Virginia Tech. Her current research interests are interdisciplinary studies of gender and engineering education research and international engineering education. She serves as Managing Editor of *Engineering Studies* and Assistant Editor of the *Global Engineering* series at Morgan & Claypool.

GARY DOWNEY

Gary Downey is Alumni Distinguished Professor of Science and Technology Studies and Affiliated Professor of Engineering Education and Women's and Gender Studies. A mechanical engineer (B.S., Lehigh University) and cultural anthropologist (Ph.D., University of Chicago), he is author of *The Machine in Me* and co-editor of *Cyborgs and Citadels*. He serves as editor of *The Engineering Studies Series* (MIT Press), *Global Engineering* series (Morgan & Claypool Publishers), and *Engineering Studies* journal (Taylor & Francis/Routledge). He is co-founder of the International Network for Engineering Studies as well as founder and co-developer of the Engineering Cultures course (ranked #2 of 190 multimedia contributions to www.globalhub.org in 2010). An ethnographic listener interested in engineering studies, he researches practices of knowledge in service.

GAYLE ELLIOTT

Gayle Elliott earned B.S. and M.S. degrees from University of Cincinnati, and is currently Associate Professor in the Division of Professional Practice. She is responsible for the University's International Co-op Program, and for placing mechanical engineering students in co-op jobs in the US. Initially part of the College of Engineering, Gayle has worked with the International Engineering Co-op Program since 1993. In 1998 she created and began working with similar programs in the Colleges of Applied Sciences, Business and the College of Design, Architecture, Art and Planning. She has extensive experience developing international exchange programs and is an active member in several international engineering education organizations and projects.

LESTER A. GERHARDT

Lester A. Gerhardt received his Bachelors degree in electrical engineering from the City College of New York (CCNY) and his Masters and Ph.D. degrees from the University of Buffalo in electrical engineering. Before beginning an academic career at Rensselaer Polytechnic Institute, he worked at Bell Aerospace Corporation on the visual simulation of space flight. His specialty is digital signal processing, emphasizing image processing, speech processing, and brain computer interfacing. He conducts sponsored research and teaching in this field, in addition to research in adaptive systems and pattern recognition, and computer integrated manufacturing. He has continuously taught each semester, and is recognized for his teaching excellence, which has included teaching at universities in Europe and Singapore. Lester is a Fellow of both the IEEE and of ASEE and holds several patents. He has served as a Board member of both privately and publically held companies, and is currently is on the Board of Directors of Capintec, Inc.

He has also been actively involved in academic administration, including positions as Department Chair, Founding Director of the Center for Manufacturing Productivity, Director of the Computer Integrated Manufacturing Program, Director of the Center for Industrial Innovation, Associate Dean of Engineering for Research and Strategy, VP of Research Administration and Finance, Dean of Engineering, and, Vice Provost and Dean of Graduate Education. His international work includes positions with NATO and the governments of Singapore, Portugal, Canada, Germany, Ireland, and the United Kingdom, in addition to co-founding the Global EEE and founding the REACH Program at Rensselaer. His honors and awards include: Inaugural Recipient of the National ASEE Research Administration Award; an honorary doctorate from the Technical University of Denmark; Distinguished Alumni Award from the State University of New York at Buffalo; and Senior Advisor to the President of the Institute of International Education (IIE).

JOHN M. GRANDIN

John M. Grandin is Professor Emeritus of German and Director Emeritus of the International Engineering Program at the University of Rhode Island, an interdisciplinary curriculum, through which students complete simultaneous degrees (BA and BS) in German, French, Spanish, or a minor in Chinese, and in an engineering discipline. Grandin has received numerous awards for his work combining languages and engineering, including the Federal Cross of Honor from the Federal Republic of Germany, the Award for Educational Innovation from ABET, and the Michael P. Malone Award for Excellence in International Education from NASULGC, the National Association of State Universities and Land Grant Colleges. He has published widely on such cross-disciplinary initiatives and has been the principle investigator for several funded projects related to the development of the International Engineering Program. Grandin also founded and organized the Annual Colloquium on International Engineering Education, bringing together university faculty and business representatives to develop a more global engineering education nationally(http://uri.edu/iep). Grandin served as associate dean and acting dean of URI's College of Arts and Sciences, and as chair of the Department of Languages. He also has published several articles and a book on Franz Kafka.

E. DAN HIRLEMAN

E. Dan Hirleman is currently professor and dean of the School of Engineering at the University of California, Merced. He earned his degrees from Purdue University, and was a National Science Foundation Graduate Fellow, a Howard Hughes Doctoral Fellow, and an Alexander von Humboldt Foundation Fellow. Hirleman received the 2006 Achievement Award from the International Network for Engineering Education and Research (INEER); ASME Fellow status in 2007; the Hon. George Brown Award for International Scientific Cooperation from the U.S. Civilian Research & Development Foundation (CRDF) in 2008, and the 2009 Charles Russ Richards Memorial Award from Pi Tau Sigma/ASME. His research is in the areas of optical sensors for surface, flow, and biohazard characterization and in global engineering education.

BRENT K. JESIEK

Brent K. Jesiek is assistant professor in Engineering Education and Electrical and Computer Engineering at Purdue University. He holds a B.S. in Electrical Engineering from Michigan Tech and M.S. and Ph.D. degrees in Science and Technology Studies from Virginia Tech. His research examines the social, historical, global, and epistemological dimensions of engineering and computing, with particular emphasis on topics related to engineering education, computer engineering, and educational technology.

JUAN LUCENA

Juan Lucena is Associate Professor at the Liberal Arts and International Studies Division (LAIS) at the Colorado School of Mines (CSM). Juan obtained a Ph.D. in Science and Technology Studies (STS) from Virginia Tech and a M.S. in STS and BS in Mechanical and Aeronautical Engineering from Rensselaer Polytechnic Institute (RPI). His book *Defending the Nation: U.S. Policy making to Create Scientists and Engineers from Sputnik to the 'War Against Terrorism'* (University Press of America, 2005) provides a comprehensive history of the education and development of scientists and engineers in the U.S. in the last five decades. In the 1990s, he researched how images of globalization shape engineering education and practice under a NSF CAREER Award titled *Global Engineers: An Ethnography of Globalization in the Education, Hiring Practices and Designs of Engineers in Europe, Latin America, and the U.S.* He has been co-investigator in projects such as *Building the Global Engineer*, aimed at developing, evaluating, and disseminating curricula on the cultural dimensions of engineering education and practice in different national contexts; *Enhancing Engineering Education through Humanitarian Ethics*, focused on researching and developing curricula at the intersection between 'humanitarianism' and 'engineering ethics'; and *Engineering and Social Justice*, aimed at finding intersections between these two apparently incommensurable fields of practice. Dr. Lucena has served as a member of key advising groups such as NSF/Sigma Xi Steering Committee on U.S. *S&E Globally Engaged S&E Workforce* and NAE's Center for Engineering Ethics and Society. He has directed the Science, Technology, and Globalization Program at Embry-Riddle Aeronautical

University and the McBride Honors Program in Public Affairs for Engineers at CSM. He has been Boeing Senior Fellow in Engineering Education at the National Academy of Engineering, Visiting Professor of Science, Engineering, and Technology Education at the Universidad de las Americas in Puebla (Mexico), and co-editor of *Engineering Studies,* the Journal of the International Network for Engineering Studies.

PHIL MCKNIGHT

Phil McKnight is Professor of German and Chair, School of Modern Languages at the Georgia Institute of Technology. He has published 9 books on East German writers, 18th century studies and satire. He initiated the Georgia Tech/TU Munich/Siemens study/internship program, and manages internships to Japan. He has received numerous major grants from the IIE, Department of Education, DAAD, Fulbright, Japan Foundation and Korean Foundation to develop applied language and intercultural studies opportunities for students at Georgia Tech. Foreign language enrollments have more than doubled since he came to Georgia Tech in 2001. Prior to Georgia Tech he was Chair of the German Department at the University of Kentucky and co-founded the Kentucky-Germany Business Council.

JAMES R. MIHELCIC

James R. Mihelcic is a Professor of Civil and Environmental Engineering and State of Florida 21st Century World Class Scholar at the University of South Florida. He directs the Peace Corps Master's International Program in Civil & Environmental Engineering (`http://cee.eng.usf.edu/peacecorps`). Dr. Mihelcic is a past president of the Association of Environmental Engineering and Science Professors (AEESP) and a member of the Environmental Protection Agency's Science Advisory Board. He is a Board Certified Environmental Engineer Member (BCEEM) and Board Trustee for the American Academy of Environmental Engineers (AAEE). Dr. Mihelcic has traveled extensively in the developing world for service and research. He is lead author for 3 textbooks: *Fundamentals of Environmental Engineering* (John Wiley & Sons, 1999); *Field Guide in Environmental Engineering for Development Workers: Water, Sanitation, Indoor Air (*ASCE Press, 2009); and, *Environmental Engineering: Fundamentals, Sustainability, Design (*John Wiley & Sons, 2010).

D. JOSEPH (JOE) MOOK

Dr. D. Joseph (Joe) Mook is currently a Program Manager in the Office of International Science and Engineering (OISE) at the National Science Foundation, where he is on leave from his permanent position as Professor of Mechanical and Aerospace Engineering (MAE) at the University at Buffalo, State University of New York. He was MAE Department Chair from 2004-2007, and Assistant Dean for International Education for the School of Engineering and Applied Sciences from 1997-2007. He is a recipient of an Alexander von Humboldt Research Fellowship (Germany), and also, a Senior Research Fellowship from the Japan Society for the Promotion of Science. He has been twice elected Chair of the Executive Committee of the Global Engineering Education Exchange (Global E3), and he has also received SUNY system-wide Chancellor's Awards for Excellence in Teaching and for Internationalization. He has been a visiting professor at Darmstadt and Hannover in Germany, at Toulouse, Troyes, and Compiegne in France, and at Chiang Mai in Thailand. His research spans topics in nonlinear optimal estimation, system identification, and controls, has resulted in approximately 100 publications, and he has supervised 13 Ph.D. and approximately 50 M.S. students to completion. He holds B.S., M.S., and Ph.D. degrees in Engineering Mechanics from Virginia Tech.

MICHAEL NUGENT

Michael Nugent is currently Director of the National Security Education Program (NSEP), which supports the Boren Scholars and Fellows Program, a major national effort to increase the quantity, diversity, and quality of the teaching and learning of subjects in the fields of foreign languages, area studies, and other international fields that are critical to the Nation's interests. Dr. Nugent also serves as Director of the Language Flagship, an NSEP program that supports professional level language learning (ILR 3 or higher) for undergraduate students of all majors at U.S colleges and universities. Along with other programs, NSEP's mission is to produce an increased pool of applicants skilled in language and culture for work in the departments and agencies of the United States Government with national security responsibilities.

Before coming to NSEP, Dr. Nugent served in a number of positions supporting international, language and cultural education at the U.S. Department of Education, including Chief of Section, Advanced Training and Research Team, managing Title VI funding of National Resource Centers, Foreign Language and Areas Studies grants, the Language Resource Centers, and the International Studies and Research Grants. He also directed the U.S.-Brazil Higher Education Consortia Program and the North America Mobility in Higher Education Program at the Fund for the Improvement of Postsecondary Education (FIPSE).

Dr. Nugent has served in policy positions as Vice President for Administration and Research at the Council for Higher Education Accreditation (CHEA) in Washington DC, and Deputy to the Chancellor for Systems Relations for Minnesota State Colleges and Universities. Author of "The Transformation of the Student Career: University study in Germany, Sweden, and the Netherlands" (RoutledgeFalmer, 2004), he remains active in the field of international higher education policy. Dr. Nugent has a Ph.D. in higher education from Pennsylvania State University. He has been a student of language and literature at universities in Germany, France, and Spain.

ALAN R. PARKINSON

Alan R. Parkinson was appointed dean of the Ira A. Fulton College of Engineering and Technology in May 2005. Before his appointment as dean, he served as associate dean and also as chair of the Mechanical Engineering department. Dr. Parkinson received his M.S. and Ph.D. degrees from the

University of Illinois and his B.S. and M.B.A. degrees from BYU. His research interests include optimization methods in engineering design, robust design, and engineering education. Dr. Parkinson received the Design Automation Award in 2003 from the American Society of Mechanical Engineers for his work in robust design. He was elected to Fellow status in the American Society of Mechanical Engineers in 2004.

LINDA PHILLIPS

Linda Phillips is a Lecturer and Patel Associate at the University of South Florida in the Civil and Environmental Department. She has a B.S. and M.S. in Civil Engineering from Michigan Technological University specializing in Construction Management. Linda has over twenty years of practical experience working as a project engineer with a large natural gas utility in Michigan and was a project manager and Vice president of Planmark Architects and Engineers, a division of SuperValu Inc., Minneapolis, Minnesota. Linda began her teaching career in 1997 at Virginia Tech and then moved to the University of Minnesota before going to Michigan Tech from 1998 – 2008, teaching classes in Project Management, Professional Practice, and Capstone Design. In 2000, at the request of her students, Linda started the International Senior Design (ISD) taking over 160 students to developing world countries to do their Capstone design projects.

MARGARET PINNELL

Dr. Margaret Pinnell is an associate professor in the Department of Mechanical and Aerospace Engineering at the University of Dayton. She teaches undergraduate and graduate materials related courses including Introduction to Materials, Materials Laboratory and Engineering Design and Appropriate Technology (ETHOS). She is the former faculty director and current associate director for ETHOS (Engineers Through Humanitarian Opportunities of Service Learning) . She has incorporated service-learning projects into her classes and laboratories since she started teaching in 2000. Her research interests include service-learning and pedagogy, K-12 outreach, biomaterials and materials testing and analysis. Prior to joining the School of Engineering, Dr. Pinnell worked at the University of Dayton Research Institute in the Structural Test Laboratory and at the Composites Branch of the Materials Laboratory at Wright Patterson Air Force Base.

ANU RAMASWAMI

Dr. Anu Ramaswami is Professor of Environmental Engineering and Director of the NSF- IGERT Program on Sustainable Urban Infrastructure at the University of Colorado Denver. Ramaswami's research spans environmental modeling, industrial ecology, sustainable infrastructure design, urban systems analysis, and integration of science and technology with policy and planning for sustainable development in communities. Her team is presently working with more than 20 cities in the US and worldwide on understanding sustainability needs, evaluating infrastructure trajectories and prioritizing actions and policies on the ground. She is also leading the development of an inter-disciplinary curriculum on "*Sustainable Infrastructures and Sustainable Communities*" for students drawn from engineering, architecture, planning, public affairs and public health. Since 1996, Ramaswami has advised more than 30 graduate students, co-authored a textbook (on Integrated Environmental Modeling), published more than 50 papers, and managed high-impact applied sustainability research and field projects in communities.

RICK VAZ

Rick Vaz is Dean of Interdisciplinary and Global Studies at Worcester Polytechnic Institute, with responsibility for a worldwide network of undergraduate research programs and an academic unit focusing on local and regional sustainability. His interests include service and experiential learning, sustainable development, and internationalizing engineering education. He held systems and design engineering positions before joining the WPI Electrical and Computer Engineering faculty in 1987.

BERND WIDDIG

Bernd Widdig is Director of the Office of International Programs and Director of the McGillycuddy-Logue Center for Undergraduate Global Studies at Boston College. Before his appointment at BC in 2007, he worked as a professor and administrator at MIT. He joined the faculty of MIT in 1989 and taught courses in German culture and literature as well as seminars on cross-cultural communication. In 1997 he started the MIT-Germany Program, a tailored international program for engineering and science students. In 2001 he was appointed Associate Director of the MIT International Science and Technology Initiative (MISTI). Bernd Widdig has published widely including *Culture and Inflation in Weimar Germany*, Berkeley: University of California Press, 2001 and *In Search of Global Engineering Excellence: Educating the Next Generation of Engineers for the Global Workplace*, Hannover 2006 (co-author). In recognition of his achievements in fostering German-American relations, he received the Cross of the Order of Merit of the Federal Republic of Germany in 2008.

Index

Printed in the United States
by Baker & Taylor Publisher Services